Springer Theses

Recognizing Outstanding Ph.D. Research

Aims and Scope

The series "Springer Theses" brings together a selection of the very best Ph.D. theses from around the world and across the physical sciences. Nominated and endorsed by two recognized specialists, each published volume has been selected for its scientific excellence and the high impact of its contents for the pertinent field of research. For greater accessibility to non-specialists, the published versions include an extended introduction, as well as a foreword by the student's supervisor explaining the special relevance of the work for the field. As a whole, the series will provide a valuable resource both for newcomers to the research fields described, and for other scientists seeking detailed background information on special questions. Finally, it provides an accredited documentation of the valuable contributions made by today's younger generation of scientists.

Theses are accepted into the series by invited nomination only and must fulfill all of the following criteria

- They must be written in good English.
- The topic should fall within the confines of Chemistry, Physics, Earth Sciences, Engineering and related interdisciplinary fields such as Materials, Nanoscience, Chemical Engineering, Complex Systems and Biophysics.
- The work reported in the thesis must represent a significant scientific advance.
- If the thesis includes previously published material, permission to reproduce this must be gained from the respective copyright holder.
- They must have been examined and passed during the 12 months prior to nomination.
- Each thesis should include a foreword by the supervisor outlining the significance of its content.
- The theses should have a clearly defined structure including an introduction accessible to scientists not expert in that particular field.

More information about this series at http://www.springer.com/series/8790

Alexander Kessel

Generation and Parametric Amplification of Few-Cycle Light Pulses at Relativistic Intensities

Doctoral Thesis accepted by
the Ludwig Maximilian University, Munich, Germany

 Springer

Author
Dr. Alexander Kessel
Laboratory for Attosecond Physics
Max-Planck-Institute for Quantum Optics
Garching, Bavaria
Germany

Supervisor
Prof. Stefan Karsch
Fakultät für Physik
Ludwig-Maximilians-Universität München
Munich
Germany

and

Max-Planck-Institute for Quantum Optics
Garching, Bavaria
Germany

ISSN 2190-5053   ISSN 2190-5061   (electronic)
Springer Theses
ISBN 978-3-030-06532-4   ISBN 978-3-319-92843-2   (eBook)
https://doi.org/10.1007/978-3-319-92843-2

Printed on acid-free paper

This Springer imprint is published by the registered company Springer International Publishing AG part of Springer Nature
The registered company address is: Gewerbestrasse 11, 6330 Cham, Switzerland

Supervisor's Foreword

Ever since strobe photography was introduced by H. Edgerton in the 1930s in order to resolve motion on timescales of milliseconds to microseconds, much shorter than the human eye can perceive, scientists have used progressively shorter flashes of light to freeze motion in fast processes. This fuels the quest for evermore intense light sources capable of delivering sufficient light for forming an image during the ever-shrinking time frame. Edgerton solved this problem at the time by using discharge flashtubes. With the invention of the laser, the intensity of light pulses took another huge leap forward. By Q-switching (Hellwarth and McClung 1962), pulse durations of nanoseconds became accessible on a routine basis, and mode-locking, invented in the early 1980s, afforded pulses in the few-femtosecond and few-cycle regime at low pulse energy. Since the early 1990s, the widespread use of the chirped-pulsed amplification (CPA) technique has heralded an increase in laser peak power by many orders of magnitude. However, as pulses are amplified, their duration is lengthened again to 10–20 fs.

Despite the remaining challenges, such laser pulses have two interesting features leading to two complementary fields of applications: Via their ultrashort duration per se they allow to trigger and measure ultrafast processes in matter and, by virtue of this short duration, their high peak power allows the study of matter under extreme conditions.

In the first field, moderately intense 20 fs laser pulses are routinely used to generate attosecond XUV pulses for ultrafast probing of matter on the shortest directly accessible timescales. Their generation involves a two-step process of nonlinear compression and frequency shifting, which is well-understood but rather inefficient and fundamentally limited in intensity. Thus, a kHz laser repetition rate is needed to achieve the required photon flux.

In the second field, Ti:sapphire lasers currently hold the world record for single-pulse peak power at 5 PW for probing matter at the extremes, but the large quantum defect requires these systems to cool down between laser shots, which limits their shot rate to once a second or lower.

Bridging the apparent contradiction between high-repetition rate, low-energy lasers for attosecond science, and high-peak-power systems for probing matter at the highest achievable intensity is the goal of the petawatt-field-synthesizer (PFS) project at the Max-Planck-Institut für Quantenoptik (MPQ). How can this be achieved?

In order to overcome the bandwidth/duration limit of available laser materials, a different amplification technique is used, called optical parametric amplification (OPA). By replacing the laser gain medium with a nonlinear crystal, it allows to transfer energy instantaneously from energetic, narrowband "pump" pulses to weak but broadband "signal" pulses. The key advantage of the OPA technique is that unlike in lasers, the amplification bandwidth depends on the thickness of the nonlinear crystal. Thin crystals can therefore solve the bandwidth issue, but in turn require short, high-intensity pump pulses in order to work efficiently. Furthermore, large crystal apertures are necessary to support the amplification of high-energy pulses which restricts the choice of nonlinear crystals to a few suitable materials. All in all, this makes for a fundamentally new system architecture that is realized within the PFS project.

An ultra-broadband front-end system provides two independent, but perfectly timed high-energy seed pulses for a picosecond pump laser and the broadband optical parametric amplifier. This synchronicity ensures that pump and signal pulses coincide during their passage through the amplifier. The pump laser is a homemade sub-picosecond, multi-terawatt CPA diode-pumped laser using Yb:YAG as the amplifying medium, and the OPA crystals, situated in vacuum vessels, are among the largest LBO crystals that can be manufactured today. Temporal compression of the OPA-amplified signal pulses down to few optical cycles of the electric field is done in an all-chirped-mirror compressor.

All these technologies were not available on a scale ready for PFS at the start of the project in 2007 and had to be developed painstakingly by two generations of Ph.D. students.

Alexander Kessel contributed to several sections of the PFS system during his Ph.D.: In a first step, he extensively reworked the scheme for ultra-broadband pulse generation in order to provide suitable seed pulses for the OPA chain. In the following, he devoted himself to the continuous development of the OPA amplifier in sync with ongoing improvements of the pump laser in which he also participated. Ultimately, he achieved output pulse energies of 45 mJ at a pulse duration of 6.4 fs and a focused intensity of 5×10^{19} W/cm^2, rivaling current record values for such OPA systems. The system now demonstrates a temporal contrast of more than eleven orders of magnitude at a rise time of one picosecond, a rate of change more than 100 times better than current state-of-the-art laser systems.

All these diverse achievements are wrapped up in a concise and insightful manner in this Ph.D. thesis. In particular, the introductory section with a mathematical description of ultrashort laser pulses and their various nonlinear interactions serves as a first-rate primer for OPCPA in a well-accessible manner. The experimental chapter gives an excellent description of the challenges of developing such a novel architecture and some of the innovative solutions to overcome them.

Garching, Germany   Prof. Stefan Karsch
January 2018

Abstract

For the generation of isolated, high-energy attosecond pulses in the extreme ultraviolet (XUV) by laser-plasma interaction on solid surfaces, there is a strong demand for light sources with exceptional properties. The key requirements are: relativistic intensities of more than 10^{19} W/cm^2, an ultrashort pulse duration with only few cycles of the electric field, and a high temporal contrast of better than 10^{10}. While state-of-the-art solid-state laser systems do reach the desired intensities, they face fundamental difficulties to generate few-cycle pulses with the required high contrast. As an alternative technique to conventional laser amplification, optical parametric chirped-pulse amplification (OPCPA) promises to fulfill all listed requirements at the same time.

In this thesis, the recent progress in the development of the Petawatt Field Synthesizer (PFS) is described, an OPCPA system that aims at generating light pulses with Joule-scale energy and ultrashort duration of 5 fs (sub-two optical cycles at 900 nm central wavelength). The octave-spanning amplification bandwidth necessary to achieve this goal is supported via the implementation of thin nonlinear crystals (LBO) for the OPCPA stages. A diode-pumped ytterbium-based amplifier chain provides the intense pump pulses for efficient parametric amplification. From the sub-picosecond pump pulse duration in combination with the instantaneous energy transfer in the OPCPA process, an excellent contrast on this timescale can be expected.

The presented work is dedicated to three major subjects: First, the generation of broadband high-energy seed pulses for parametric amplification is discussed. To this end, different schemes were set up and tested, where special effort was made to produce pulses with a smooth spectral intensity and phase to avoid the aggravation of distortions during later amplification.

Second, the parametric amplification of the stretched seed pulses in two OPCPA stages from few µJ to 1 mJ after the first stage and more than 50 mJ after the second stage is presented. By all-chirped-mirror compression of the amplified pulses, a pulse duration of 6.4 fs was achieved, resulting in an effective peak power of 4.9 TW and a peak intensity of 4.5×10^{19} W/cm^2 after focusing. The temporal contrast of the pulses was measured to be better than 10^{11} starting from 1 ps before the main peak, which demonstrates the potential of OPCPA systems that employ short pump pulses. At this performance, the PFS ranks among the most powerful few-cycle light sources existing today and is currently used for first high-harmonic generation (HHG) experiments in our laboratory.

Finally, preparations were made for an upgrade of the current system which is expected to boost the output power by more than an order of magnitude toward the 100 TW regime. The necessary upscaling of beam diameters for this step required the determination of optimal parameters for the large nonlinear crystals to be purchased. Furthermore, a concept was developed to match the pulse fronts of pump and signal beams in the non-collinear OPA stages.

Publications Related to this Work

A. Kessel, C. Skrobol, S. Klingebiel, C. Wandt, I. Ahmad, S. A. Trushin, Zs. Major, F. Krausz, and S. Karsch, "Generation and optical parametric amplification of near-IR, few-cycle light pulses", *CLEO: 2014 OSA Technical Digest (online)*, paper SM3I.3, (2014), https://doi.org/10.1364/CLEO_SI.2014.SM3I.3.
→ Contributions by the author: performing the experiments, data analysis, writing the manuscript

C. Wandt, S. Klingebiel, S. Keppler, M. Hornung, C. Skrobol, A. Kessel, S. A. Trushin, Zs. Major, J. Hein, M. C. Kaluza, F. Krausz, and S. Karsch, "Development of a Joule-class Yb:YAG amplifier and its implementation in a CPA system generating 1TW pulses", Laser Photonic Rev. **881**, (2014), https://doi.org/10.1002/lpor.201400040.
→ Contributions by the author: providing the frontend pump seed pulses for the experiments, discussion of the results, proofreading of the manuscript

A. Kessel, S. A. Trushin, N. Karpowicz, C. Skrobol, S. Klingebiel, C. Wandt, and S. Karsch, "Generation of multi-octave spanning high-energy pulses by cascaded nonlinear processes in BBO", Opt. Express **24**, (2016), https://doi.org/10.1364/OE.24.005628.
→ Contributions by the author: performing the experiments, data analysis, development of the code for the simulation, writing the manuscript

A. Kessel, V. E. Leshchenko, M. Küger, O. Lysov, A. Münzer, A. Weigel, V. Pervak, M. Trubetskov, S. A. Trushin, Zs. Major, F. Krausz, and S. Karsch, "Broadband Picosecond-Pumped OPCPA Delivering 5 TW, Sub-7 fs Pulses with Excellent Temporal Contrast", accepted contribution to *CLEO Europe 2017*
→ Contributions by the author: performing the experiments, data analysis, writing the manuscript

A. Kessel, V. E. Leshchenko, O. Jahn, M. Krüger, A. Münzer, A. Schwarz, V. Pervak, M. Trubetskov, S. A. Trushin, F. Krausz, Zs. Major, S. Karsch, "Relativistic few-cycle pulses with high contrast from picosecond-pumped OPCPA", Optica **5**, (2018), https://doi.org/10.1364/OPTICA.5.000434

Other Publications at MPQ

→ The following publications have been made by the group of Prof. Kling at the MPQ. The author of this work has contributed to these publications by providing the laser pulses from the PFS frontend system (comprising the Ti:Sa oscillator and Femtopower amplifier) and helping with pulse compression and CEP stabilization.

H. Li, A. Alnaser, X.-M. Tong, K. Betsch, M. Kübel, T. Pischke, B. Förg, J. Schötz, F. Sümann, S. Zherebtsov, B. Bergues, A. Kessel, S. A. Trushin, A. M. Azzeer, and M. F. Kling, "Intensity dependence of the attosecond control of the dissociative ionization of D_2", J. Phys. B **47**, 124020 (2014)

F. SÜMANN, L. SEIFFERT, S. ZHEREBTSOV, V. MONDES, J. STIERLE, M. ARBEITER, J. PLENGE, P. RUPP, C. PELTZ, **A. KESSEL**, S. A. TRUSHIN, B. AHN, D. KIM, C. GRAF, E. RÜHL, M. F. KLING, AND T. FENNEL, "Field propagation-induced directionality of carrier-envelope phase-controlled photoemission from nanospheres", Nat. Commun. **6**, 7944 (2015)

H. LI, B. MIGNOLET, G. WACHTER, S. SKRUSZEWICZ, S. ZHEREBTSOV, F. SÜMANN, **A. KESSEL**, S. A. TRUSHIN, N. G. KLING, M. KÜBEL, B. AHN, D. KIM, I. BEN-ITZHAK, C. COCKE, T. FENNEL, J. TIGGESBÄUMKER, K.-H. MEIWES-BROER, C. LEMELL, J. BURGDÖRFER, R. D. LEVINE, F. REMACLE, AND M. F. KLING, "Coherent electronic wave packet motion in C_{60} controlled by the waveform and polarization of few-cycle laser fields", Phys. Rev. Lett. **114**, 123004 (2015)

H. LI, X.-M. TONG, N. SCHIRMEL, G. URBASCH, K. BETSCH, S. ZHEREBTSOV, F. SÜMANN, **A. KESSEL**, S. A. TRUSHIN, G. G. PAULUS, K. M. WEITZEL, AND M. F. KLING, "Intensity dependence of the dissociative ionization of DCl in few-cycle laser fields", J. Phys. B **49**, 015601 (2016)

H. LI, N. KLING, B. FÖRG, J. STIERLE, **A. KESSEL**, S. A. TRUSHIN, M. F. KLING, AND S. KAZIANNIS, "Carrier-envelope phase dependence of the directional fragmentation and hydrogen migration in toluene in few-cycle laser fields", Structural Dynamics **3**, 043206 (2016)

B. FÖRG, J. SCHÖTZ, F. SÜMANN, M. FÖRSTER, M. KRÜGER, B. AHN, W. OKELL, K. WINTERSPERGER, S. ZHEREBTSOV, A. GUGGENMOS, V. PERVAK, **A. KESSEL**, S. A. TRUSHIN, A. M. AZZEER, M. I. STOCKMAN, D. KIM, F. KRAUSZ, P. HOMMELHOFF, AND M. F. KLING, "Attosecond nanoscale near-field sampling", Nat. Comm. **7**, 11717 (2016)

P. RUPP, L. SEIFFERT, C. LIU, F. SÜMANN, B. AHN, B. FÖRG, C. SCHÄFER, M. GALLEI, V. MONDES, **A. KESSEL**, S. A. TRUSHIN, G. GRAF, E. RÜHL, J. LEE, M. KIM, D. KIM, T. FENNEL, M. F. KLING, AND S. ZHEREBTSOV, "Quenching of material dependence in few-cycle driven electron acceleration from nanoparticles under many-particle charge interaction", J. Mod. Opt. **64**, 995 (2016)

Publications at Other Institutes

Z. YANG, **A. KESSEL** AND G. HÄUSLER, "Better 3D inspection with structured illumination: signal formation and precision", Appl. Opt. **54** (22), 6652–6660 (2015)

→ Contributions by the author: joint development of the experimental setup in the course of the diploma work, theoretical analysis of the signal formation and the precision limits

Acknowledgements

Probably most fellow students will agree with me that doing your Ph.D. sometimes feels like running a marathon with an ever-moving finish line. Therefore, I would like to express my gratitude to all those who helped me to finally cross it.

First of all, I would like to thank:

- my supervisor Prof. Stefan Karsch for his guidance and advice, for giving me the freedom to explore things on my own, and for his support in solving the never-ending physical puzzles in the laboratory.
- Prof. Ferenc Krausz for giving me the opportunity to work in his renowned group at the Laboratory for Attosecond Physics (LAP) with its great infrastructure.
- Prof. Malte Kaluza for kindly reviewing this work as a second supervisor.
- Dr. Zsuzsanna Major, for having an open ear for all scientific and non-scientific matters and for her organizational skills and the proofreading of this work.

Starting my Ph.D. at the MPQ few years after the launch of the PFS project, I had the pleasure to work together with two generations of Ph.D. students and young postdocs. So my gratitude goes equally to the original series with Christoph Skrobol, Sandro Klingebiel, and Christoph Wandt and to the next generation with Vyacheslav Leshchenko, Alexander Schwarz, Mathias Krüger, Olga Lysov, and Andreas Münzer. Thanks for your support! It was a great time with you inside and outside the laboratory!

Special thanks go to Sergei Trushin, our senior postdoc, for introducing me to the front-end system, his daily support with starting up the laser, the many lively discussions, and his patience.

Thanks also to the master and bachelor students in our group, Susanne Eder, Alexander Kastner, and Kilian Fritsch, for their great work and the pleasant time together.

Many thanks to the ATLAS team with Johannes Wenz, Kostja Khrennikov, Matthias Heigoldt, Max Gilljohann, Mathias Hüther, and Hao Ding for answering my ignorant plasma-physics questions and for the entertaining skiing trips and Wiesn visits.

A big thank you to my great office mates Harshit Lakhotia and Daniel Cardenas. It was nice to share a room with you guys!

Thanks also to the AS5 team with Frederik Süßmann, Benjamin Förg, Johannes Schötz, William Okell, Hui Li, Philipp Rupp, and Sergey Zherebtsov for the good cooperation in the laboratory when sharing one laser front end on two floors.

Thanks as well to the (former) AS1 team: Tran Trung Lu, Mohammed Hassan, and Antoine Moulet. It was fun to share the laboratory with you, because of all the laughter and also the occasionally quite interesting music coming from behind the laser curtain.

A big thank you to all the other inhabitants of the attoworld for the marvelous working atmosphere, the canteen and hallway discussions on all kind of subjects, and the generous exchange of laboratory equipment. This goes in particular to Hanieh Fattahi (minister of communication), Olga Razskazovskaya, Sabine Keiber and Annkatrin Sommer, to Clemens Jakubeit, Matthew Weidman, Daniel Rivas, Alexander Guggenmos, Vladislav Yakovlev, Boris Bergues, Shao-Wei Chou, and Tim Paasch-Colberg.

A special thanks to Nick Karpowicz for his unshakable friendliness and for sharing his tremendous physical knowledge and his coding skills with everyone interested.

Thanks to Prof. Laszlo Veisz for his valuable comments and advice, especially considering the fact that the PFS tries to surpass his beloved (and excellent) LWS20 system. Thanks for this openness.

Many thanks to Volodymyr Pervak and Michael Trubetskov for the design and production of several types of their famous chirped mirrors.

Thanks also to Alexander Weigel for providing the Tundra device and his support for the contrast measurements.

A big thank you to the IMPRS-APS coordinators Prof. Matthias Kling, Nick Karpowicz, and Hanieh Fattahi and most of all to Mrs Wild for the fantastic annual meetings and workshops at Schloss Ringberg, in Vienna and Wildbad Kreuth.

Ein großes System wie der PFS mit unzähligen Komponenten kann nicht allein von Doktoranden und Postdocs aufgebaut und instand gehalten werden. Deswegen geht ein großer Dank für Ihre Hilfe an das Techniker-Team am MPQ: Harald Haas, Alois Böswald, Martin Fischer, Anton Horn, Tobias Kleinhenz und Martin Triphahn.

Ebenso möchte ich mich beim Werkstatt-Team um Michael Rogg und Martin Gruber für die kundige Beratung und die schnelle Fertigung der (öfter mal kurzfristig) benötigten Teile bedanken.

Ein Dankeschön auch an Franziska Hoss, Tanya Bergues, Amu Shrestha ("the excellent") und Nils Haag für die nützlichen Infos und unterhaltsamen Mittagspausen.

Dank auch an Wolfgang Huber dafür, immer einen schon ausgemusterten PC für uns auf Lager zu haben.

Vielen Dank an die Einkaufsabteilung, insbesondere Carina Steinkrebs, für die Unterstützung bei komplizierten Bestellungen.

Zu guter Letzt gilt mein Dank natürlich Ka, der besten Freundin von allen, meinen Freunden und meiner Familie. Vielen Dank für die Unterstützung über all die Jahre!

Contents

1   **Introduction**   1
    1.1   Short Light Pulses—Motivation and Historical Background   1
    1.2   The PFS Project   4
    1.3   Structure of the Thesis   6
    References   9

2   **Fundamentals**   13
    2.1   Mathematical Description of Ultrashort Light Pulses   13
        2.1.1   Derivation of the First-Order Propagation Equation   13
        2.1.2   Spatial Description and Diffraction of Light Pulses   17
        2.1.3   Temporal and Spectral Description of Light Pulses   20
        2.1.4   Dispersion and Absorption   22
        2.1.5   Chirped Pulse Amplification and Dispersion Control   24
    2.2   Nonlinear Processes   27
        2.2.1   Second-Order Processes   27
        2.2.2   Third-Order Processes   32
    2.3   Nonlinear Media   35
        2.3.1   Isotropic Media   36
        2.3.2   Crystals   37
        2.3.3   Critical Phase Matching   40
        2.3.4   Nonlinear Susceptibility Tensors   43
    2.4   Numerical Simulations   44
    2.5   Temporal Characterization of Ultrashort Light Pulses   47
        2.5.1   Cross-Correlation FROG   49
        2.5.2   Single-Shot Second-Harmonic Generation FROG   50
        2.5.3   Transient-Grating FROG   53
    2.6   Summary   54
    References   55

3   The Petawatt Field Synthesizer System 59
   3.1   Frontend and Initial Seed Generation Approach 59
   3.2   The Pump Chain 62
   3.3   The OPCPA System 66
       3.3.1   Previous Experiments........................... 66
       3.3.2   The Current System............................ 68
       3.3.3   Timing Jitter 68
   References 71

4   Seed Generation Schemes 75
   4.1   Idler Generation 75
       4.1.1   Experimental Realization 76
       4.1.2   Compensation of the Angular Chirp 79
       4.1.3   Possible Improvements of the Setup 83
   4.2   Cross-Polarized Wave Generation 84
   4.3   Cascaded Difference-Frequency Generation 86
       4.3.1   Experimental Setup and Findings 87
       4.3.2   Simulation................................... 90
       4.3.3   Discussion and Conclusions 92
   References 94

5   OPCPA Experiments with Two OPA Stages 97
   5.1   Performance of Alternative Seed Generation Schemes 97
   5.2   High-Energy OPCPA Experiments 101
       5.2.1   Pump 101
       5.2.2   Signal..................................... 104
       5.2.3   OPCPA.................................... 108
       5.2.4   Summary 117
   References 118

6   Preparations for a Third OPA Stage 119
   6.1   General Layout....................................... 119
   6.2   Evaluation of Nonlinear Crystals 121
       6.2.1   Damage Threshold Measurements.................. 122
       6.2.2   Determination of Optimal Thickness................ 123
   6.3   Pulse Front Matching 129
       6.3.1   PFT Control with Transmission Gratings.............. 131
       6.3.2   PFT Control by Adjustment of the Pump Compressor 135
   References 140

7   Summary and Outlook 143
   7.1   Summary 143
   7.2   High Peak-Power Systems Worldwide 144
   7.3   Current Works and Outlook 146
   References 147

Appendix A: Supplementary Calculations and Experiments 151

Appendix B: Data Archiving 165

Acronyms

AOM	Acousto-optic modulator
BBO	Beta barium borate, β-BaB_2O_4
CCD	Charge-coupled device
CEP	Carrier envelope phase
CPA	Chirped-pulse amplification
DFG	Difference-frequency generation
DKDP	Potassium dideuterium phosphate, KD_2PO_4
FOD	Fourth-order dispersion
FOPE	First-order propagation equation
FROG	Frequency-resolved optical gating
FWHM	Full width at half maximum
GD	Group delay
GDD	Group-delay dispersion
HCF	Hollow core fiber
HHG	High-harmonic generation
IR	Infrared
LBO	Lithium triborate, LiB_3O_5
LC-SLM	Liquid-crystal spatial light modulator
MCP	Multi-channel plate detector
Nd:YAG	Neodymium-doped yttrium aluminum garnet, $Nd^{3+}:Y_3Al_5O_{12}$
NOPA	Non-collinear optical parametric amplification
OPA	Optical parametric amplification
OPCPA	Optical parametric chirped-pulse amplification
PDE	Partial differential equation
PFS	Petawatt field synthesizer
ROM	Relativistic oscillating mirror
SFG	Sum-frequency generation
SHG	Second-harmonic generation
SHHG	High-harmonic generation on solid surfaces
SI	Spectral interferometry

SPIDER	Spectral phase interferometry for direct electric-field reconstruction
SPM	Self-phase modulation
SRSI	Self-referenced spectral interferometry
SS-SHG-FROG	Single-shot, SHG-based frequency-resolved optical gating
SWIR	Short-wavelength infrared
TFP	Thin-film polarizer
TG-FROG	Transient-grating frequency-resolved optical gating
THG	Third-harmonic generation
Ti:Sa	Titanium-sapphire, Ti:Al_2O_3
TOD	Third-order dispersion
UV	Ultraviolet
XFROG	Cross-correlation frequency-resolved optical gating
XPM	Cross-phase modulation
XPW	Cross-polarized wave generation
XUV	Extreme ultraviolet
YAG	Yttrium aluminum garnet, $Y_3Al_5O_{12}$
Yb:YAG	ytterbium-doped yttrium aluminum garnet, Yb^{3+}:$Y_3Al_5O_{12}$

Chapter 1
Introduction

1.1 Short Light Pulses—Motivation and Historical Background

Light plays an essential role in our everyday life. Its detection via the eye provides us with continuous information about the objects and dynamics in our surroundings. For scientists, it has been a strong motivation to surpass the capabilities of the eye by technological means in order to gain insights into natures' structures and processes that would be otherwise too small, too fast or too weak to be observed. Over the past centuries this desire has led to great innovations on the fields of photography, microscopy, astronomy and many others.

The experimental realization of the *laser* ("light amplification by stimulated emission of radiation") in 1960 [1] constitutes a major step regarding this goal and enabled a variety of new techniques and applications in the last decades such as optical tweezers [2], frequency combs [3], or highly precise interferometers [4].

One of the most interesting consequences of the laser principle is the ability to concentrate energy into very short pulses by the coherent superposition of light waves. For the study of fast physical processes this is a valuable property as it allows to resolve these processes on a time scale shorter than their temporal constants. Hence, with picosecond ($1 \, \text{ps} = 10^{-12}$ s) laser pulses generated in the 1970s, for the first time the vibrational dynamics of molecules could be observed [5]. The development of femtosecond ($1 \, \text{fs} = 10^{-15}$ s) lasers in the 1980s opened the door to the measurement (and control) of atomic motion and chemical reactions [6].

Using *titanium-sapphire (Ti:Sa)* as a gain medium, conventional laser schemes have been optimized to their limits in terms of spectral bandwidth and pulse duration. Originally developed in the late 1980s and early 1990s [7, 8], Ti:Sa-based laser systems until today constitute the main work horse in many laser labs and may provide $\sim$5 fs pulses on a nanojoule-level from oscillators [9]. One of the main obstacles on the route towards even shorter pulses is the fluorescence emission bandwidth of the laser gain medium [10] that limits the maximum spectral width to less than one

© Springer International Publishing AG, part of Springer Nature 2018
A. Kessel, *Generation and Parametric Amplification of Few-Cycle Light Pulses at Relativistic Intensities*, Springer Theses, https://doi.org/10.1007/978-3-319-92843-2_1

optical octave. Moreover, this physical property also leads to a spectral narrowing of the pulses during further amplification ("gain narrowing"): as a consequence, high-energy pulses on the millijoule- to joule-level from Ti:Sa power amplifiers typically feature pulse durations only in the range of 20–30 fs [11, 12].

Hence, for the next big step towards ever shorter pulses—entering the attosecond (1as = 10^{-18} s) regime which allows the observation and control of electronic motion—a fundamentally new approach was required.

Already in 1961, one year after the first demonstration of the laser, Franken et al. exploited the temporal confinement of energy in short laser pulses to achieve electric field strengths of $\sim 10^7$ V/m, approaching the binding forces of electrons in matter [13]. Triggering in this way a nonlinear response in crystalline quartz, the second harmonic of the fundamental laser pulse was generated (SHG), giving birth to the field of *nonlinear optics*.

Thanks to advances in laser development with ever increasing peak intensities, multiple other nonlinear effects were observed in the following decades. Of particular interest for the generation of ultrashort pulses is the nonlinear spectral broadening of laser pulses in gas-filled capillaries [14]. By this technique, amplified Ti:Sa pulses with mJ energy can be compressed to sub-4 fs (about two optical cycles) [15]. While not being substantially shorter than Ti:Sa oscillator pulses ($\sim$5 fs), the boost in pulse energy allows to drive another nonlinear process: the *generation of high harmonics* (*HHG*) [16]. This process can be triggered by focusing an intense laser pulse into a gas jet which results in the ionization of the atoms. The free electrons follow the oscillations of the electric field of the pulse and generate for each re-collision with the quasi-static parent ions a short burst of extreme ultraviolet (XUV) radiation, forming a train of attosecond pulses. Due to the high nonlinearity of the process, this train can be confined to a single attosecond pulse when using few-cycle driving pulses such as the mentioned spectrally broadened Ti:Sa output ("amplitude gating") [17]. At record pulse durations of less than 100 as [18], these isolated attosecond pulses provide until today an unrivaled temporal resolution. A common application here is the pump-probe technique where a fs-pulse triggers electronic motion in solids ("pump") and the as-pulse serves as an ultra-fast camera to resolve the response ("probe") [19–21].

To achieve ultimate temporal resolution and enable an even broader range of applications, it is desirable to use the attosecond XUV pulses not only as a probe but also as the pump. This in turn requires an increase of the pulse energy by orders of magnitude which proved to be difficult to accomplish with the described HHG technique in gas: As the peak intensity of the driving pulse is limited by the ionization threshold of the target gas medium, a loose-focusing geometry is obligatory for high-energy laser systems. This geometry becomes impractically large (several meters) when the few-cycle driving pulses approach energies of tens of mJ [22–25].

A method that circumvents the intensity limitation of HHG in gas and at the same time exhibits significantly higher conversion efficiencies, is the generation of high harmonics on plasma targets [26–30]. In this process an intense light pulse interacts with a solid surface (*SHHG*), ionizes it and creates a nearly step-like vacuum-plasma interface. Under the effect of the ponderomotive force of the incident pulse and the restoring Coulomb force of the ionized background, the free electrons perform oscillations perpendicular to the surface, forming a plasma mirror that moves at relativistic speed. The incident light pulse that creates this oscillating mirror (ROM), at the same time is reflected from it. Owing to the relativistic oscillations of the mirror, it gets Doppler-shifted resulting in high-harmonic radiation in the deep XUV and even soft X-ray region. In the temporal domain, this corresponds to the generation of one attosecond pulse per cycle of the electric field where once again amplitude gating can be used to select an isolated pulse.

To perform SHHG experiments, relativistic intensities larger than 1.37×10^{18} W/cm^2 $\times \lambda_c^2$ [µm] are required on target, where λ_c is the central wavelength of the driving pulse. For few-cycle pulses with $\lambda_c \sim 1$ µm and a typical focal spot size of few µm^2, this calls for pulse energies of at least the order of mJ. For an efficient operation of the process and for the generation of high harmonics up to the water window (>280 eV), however, rather hundreds of mJ are necessary [27].

Even though conventional laser systems exist that provide pulses with Petawatt peak power and strongly relativistic intensities [31, 32], their long pulse duration prevents the generation of isolated attosecond pulses by SHHG. The aforementioned temporal compression to the few-cycle regime by nonlinear spectral broadening does not help here as the technique is limited in energy: so far only the generation of sub-two-cycle pulses at $\sim$3 mJ pulse energy [33] and sub-four-cycle pulses at $\sim$10 mJ [34] have been demonstrated. A recently suggested scheme for upscaling the concept to the joule-level by spectral broadening in thin foils still has to show its experimental feasibility [35, 36].

Besides the limitation in pulse duration, conventional lasers face another issue when used for plasma experiments, namely the imperfect temporal contrast: Essentially all high-energy laser pulses exhibit an amplified spontaneous emission (ASE) background and a coherent pedestal from pulse compression after chirped-pulse amplification (see for example [37]). As the intensity of these artifacts is typically of the order of 10^{-9} relative to the intensity of the main pulse, they may well exceed the threshold intensity for plasma formation of $\sim$10^{10} W/cm^2. Due to the long temporal extent of the pedestals, this occurs already tens (coherent pedestal) or hundreds (ASE) of picoseconds before the main pulse arrives, giving the generated plasma time to expand. Thus, in the case of SHHG the otherwise clean plasma-vacuum interface created by the main pulse is critically distorted, impairing the generation of high harmonics.

Next Generation Relativistic Light Sources

Mainly for the reason of contrast improvement, several large-scale systems nowadays combine the standard laser amplification with optical parametric amplification (OPA) [38–40]. This technique exploits the second-order nonlinearity of birefringent crystals to transfer energy from an energetic, narrowband "pump" beam to a weak, but broadband "signal" beam [41, 42]. A third beam—called "idler"—is generated in this three-wave-mixing process and guarantees conservation of energy and momentum. Being an instantaneous process, no energy is stored in the nonlinear crystal and therefore parasitic effects are limited to the duration of the pump pulse. This property makes OPA schemes an ideal frontend for systems with high contrast requirements.

Going one step further, conventional laser amplifiers can be entirely replaced by parametric amplifiers, which has already been demonstrated on a single shot basis for high energies (35 J, 84 fs) [43]. The main motivation behind this step is the ability of the OPA technique to amplify ultrashort pulses as the bandwidth depends only on the phase-mismatch between pump, signal and idler that is aggregated by propagation of the pulses inside the nonlinear crystal (assuming no strong absorption of spectral components in the medium). At multi-TW peak power, this capability has been impressively shown by the LWS20 system delivering 4.5 fs pulses with up to 80 mJ energy at a 10 Hz repetition rate [44].

This thesis is dedicated to the continued development of a light source that is planned to take the concept of parametric ultrashort pulse amplification to the Petawatt-level.

1.2 The PFS Project

The Petawatt Field Synthesizer (PFS) at the Max-Planck-Institut für Quantenoptik (MPQ) has been designed to deliver sub-two-cycle, high-contrast light pulses for the generation of high harmonics on solid surfaces as the main application [45, 46]. To reach the ambitious target parameters—a pulse duration of 5 fs, pulse energies of up to 3 J and a repetition rate of 10 Hz—the concept of this fully OPA-based system contains several key features, which will be briefly explained in the following.

In order to scale OPA systems to high pulse energies, the amplification is distributed to several consecutive OPA stages similar to the amplifier chains in conventional laser systems. The number of required stages depends on the achievable gain per stage which in turn is determined by experimental conditions such as the amplification bandwidth or the type of nonlinear crystal. To reach the PFS target energy of 3 J, the last stages have to be pumped by *multi-joule pump pulses* requiring *large nonlinear crystals* and beam diameters of the order of several centimeters to keep the fluences below the damage threshold of the medium.

At the time of design of the system in 2007, potassium dideuterium phosphate KD_2PO_4 (DKDP) was the only suitable nonlinear crystal type available at this size. Due to its comparatively weak nonlinearity, high pump intensities are necessary to maintain an efficient parametric amplification—at first glance a contradiction to the prior condition of low fluences to prevent optical damage. The solution is to use *short pump pulses* at ~1 ps pulse duration: Owing to the $\sqrt{\tau}$-scaling of the damage threshold fluence for pulse durations τ in the nanosecond to picosecond range [47], a safe operation of the OPA stages at intensities of the order of 100 GW/cm^2 is possible.

In the first (small) OPA stages, DKDP was later replaced by lithium triborate (LBO) owing to its superior nonlinear properties. As a result, in the design of the system the total number of required stages could be reduced from 8 to 5 [48, 49].

To reach the desired amplification bandwidth of about one optical octave, the phase-mismatch between the three interacting waves in the OPA process is kept small by employing *thin nonlinear crystals*. Due to the shorter interaction length, choosing this solution usually comes at the price of a reduced OPA gain. The high-intensity pump pulses, however, mostly compensate for this effect. Overall, this ansatz constitutes a notable difference to high-energy OPA systems using longer pump pulses and thicker crystals where two-color pumping is required to achieve a similar bandwidth [50].

For an efficient transfer of energy from the pump pulses to the signal, the temporal overlap between both pulses during amplification should be as large as possible [51]. This is accomplished by introducing a chirp in the 5 fs signal pulses before amplification in order to stretch them to the ∼1 ps pulse duration of the pump. After amplification, the signal pulses are re-compressed to achieve highest peak-powers.

In analogy to the chirped pulse amplification (CPA) technique known from conventional laser amplifiers [52] this method is termed *optical parametric chirped pulse amplification (OPCPA)*. Compared to typical CPA schemes, the required chirp is orders of magnitude smaller—owing to the broad signal bandwidth and the short pump pulse duration. On the other hand, the 5 fs target duration for the amplified signal pulses poses high requirements to the compensation of higher order dispersion.

As mentioned before, an excellent temporal contrast is vital for HHG experiments on solid surfaces—the main application of the PFS. Considering the potential peak intensity of up to 10^{22} W/cm^2 in the fully realized version of the system, a contrast of better than 10^{12} has to be provided to stay below the plasma formation threshold of ∼10^{10} W/cm^2. The PFS concept not only promises to reach and surpass such contrast ratios owing to the intrinsic temporal filtering via the OPA technique but is expected to do so even on a 1 ps time scale due to the short pump pulses, thus outperforming OPA systems with longer pump pulse durations.

For the experimental realization of the PFS concept, three *main challenges* have to be mastered:

(I) The generation of 1 ps *pump* pulses featuring energies of more than 10 J in the fundamental beam at a repetition rate of 10 Hz. Not being commercially available, a system delivering such pulses based on ytterbium-doped gain media (1030 nm central wavelength) had to be newly developed at the MPQ.

(II) The generation of high-energy, broadband *signal ("seed")* pulses for the OPA chain from a commercial Ti:Sa frontend. The required spectral range of 700–1400 nm for these pulses is defined by the frequency-doubled pump wavelength of 515 nm and the phase-matching properties of the DKPD and LBO crystals.

(III) The combination of pump and signal for *parametric amplification*. This involves the installation and optimization of the OPA system as well as the dispersion management for the signal pulses, i.e. the temporal stretching before and re-compression after amplification.

Regarding all three pillars significant achievements have been attained by co-workers, namely (in approximately historical order) the installation of the PFS frontend including a preliminary scheme for broadband seed generation [53, 54], the development of an Yb-based CPA pump amplifier chain delivering up to 0.8 J compressed energy at a pulse duration of ∼800 fs [55–59] as well as first proof-of-principle OPA experiments at a reduced pump energy [48, 49].

Building on these earlier works, we will describe in the following chapters new developments, modifications and improvements to the PFS system, with the main focus on seed generation (pillar II) and OPA development (pillar III). The overall goal of these measures is to reach for the first time relativistic intensities exceeding 10^{19} W/cm^2 that allow to perform SHHG experiments.

1.3   Structure of the Thesis

Chapter 2—Fundamentals
This chapter provides an introduction into the mathematical description of ultrashort light pulses. The linear and nonlinear physical effects relevant to this work are discussed and experimental tools for temporal characterization are presented.

Chapter 3—The Petawatt Field Synthesizer System
This chapter gives an overview over the PFS system, summarizes previous works and explains new or modified parts of the system. In particular the initial seed generation scheme is described which delivered the broadband pulses for the OPCPA chain by spectral broadening in two cascaded hollow-core fibers (HCF). The shortcomings of this scheme provide the motivation for Chap. 4.

Main modifications to the system:
- The replacement of the master oscillator with an updated version simplifies the pump chain owing to an additional 1030 nm output.
- The rebuild of the pump compressor in vacuum slightly reduces the timing jitter and allows the compression of high-energy pump pulses (potentially >10 J).
- The rebuild of the currently last pump amplifier enables a stable long-term operation and improves the maintainability of the setup, the beam profile and the output energy.

Chapter 4—Seed Generation Schemes
In this chapter the principle ideas and experimental realizations of alternative schemes for the generation of broadband seed/signal pulses for the OPCPA chain are explained. In total three different schemes are tested that use a Ti:Sa amplifier (1.5 mJ, 30 fs, 1 kHz) as a common frontend.

4.1   Idler Generation
In this scheme the Ti:Sa pulses are split into two channels, where one channel is frequency-doubled and the other broadened in a HCF. By recombination in a non-collinear OPA stage, a broadband idler is produced. Theoretical considerations and experiments are presented, aiming at the compensation of the intrinsic angular chirp of this idler.

Main results:

- The generated idler pulses feature energies of up to $15\,\mu J$ and a spectral range of 680–1400 nm.
- The experimentally achievable compensation of the angular chirp is not sufficient to allow the use of the generated pulses as a seed.

4.2 Cross-Polarized Wave Generation

Reusing the original seed generation scheme consisting of two HCFs, the broadband pulses are temporally gated and spectrally smoothed in a cross-polarized wave (XPW) generation stage.

Main results:

- $4\,\mu J$ pulses are generated that feature a smooth spectrum spanning 680–1600 nm with an almost Gaussian shape.
- The reliability of the setup makes it the scheme of choice for daily operation as an OPA seed source.

4.3 Cascaded Difference-Frequency Generation

After pulse compression of the Ti:Sa pulses in one HCF, strong broadening into the infrared is achieved by cascaded second-order intra-pulse interactions in a BBO crystal.

Main results:

- Generation of $10\,\mu J$ pulses with extremely broad spectra from 600 nm to more than 2400 nm ($\sim 8\,\mu J$ in the relevant range of 700 nm to 1400 nm).
- Being conceptually simple and scalable in energy, the scheme is recommended as the future seed source for the PFS.

Chapter 5—OPCPA Experiments with Two OPA Stages

This chapter describes the OPCPA experiments carried out in vacuum on two OPA stages with LBO crystals.

5.1 Performance of Alternative Seed Generation Schemes

The seed pulses from the idler generation and the XPW generation scheme are parametrically amplified and compared to prior experiments with the initial seed generation scheme. Furthermore, the temporal compressibility of the amplified XPW pulses is tested in air.

Main results:

- Compared to the initial seed generation scheme, both alternative schemes yield (at a comparable gain) significantly less modulated spectra of the amplified pulses promising better compressibility.
- Seeding with the XPW scheme, energies of $180\,\mu J$ and 9.8 mJ are achieved after the first and second OPA stage respectively (4.5 and 81 mJ pump energy at 1 Hz).

- Test compression of the amplified pulses with chirped mirrors in air results in a pulse duration of 7 fs.

5.2 High-Energy OPCPA Experiments

For the first time the full 1.2 J fundamental pulse energy available from the last pump amplifier are used for OPA experiments at 10 Hz

Main results:

- The pump SHG efficiency is improved from previously 45% to 55% by the implementation of a fast-switching Pockels cell in the amplifier chain to suppress amplified spontaneous emission. This results in frequency-doubled pump pulse energies of 14 and 400 mJ respectively for the two OPA stages.
- The parametric amplification of the broadband seed pulses generated by the XPW scheme yields pulse energies of 1.0 mJ after the first and 53 mJ after the second OPA stage.
- These amplified signal pulses are compressed to approximately 6.4 fs (sub-2.5-optical cycles) with chirped mirrors in vacuum, resulting in a peak power of 4.9 TW.
- The temporal contrast of the amplified signal pulses is measured to be better than 10^{11} on a 1 ps time scale.

Chapter 6—Preparations for a Third OPA Stage

In this chapter the preparations for the next upgrade of the PFS system are presented. This includes the basic design considerations for the system, experimental tests of nonlinear crystals and an examination of potential schemes to match the pulse fronts of pump and signal in the upscaled system.

6.1 General Layout The designed layout for the system upgrade is discussed.

Main planned modifications:

- The existing system will be extended with a third (using pump recycling possibly even with a fourth) OPA stage.
- A new pump amplifier is planned to deliver the required energy of 10 J in the fundamental beam. After splitting this beam into a 9 J and a 1 J arm, the pulses will be individually compressed and frequency-doubled, providing 4 J and 400 mJ + 15 mJ pump energy for the OPA stages.

6.2 Evaluation of Nonlinear Crystals

Possible options regarding the nonlinear crystal type (DKDP or LBO) to be used for pump SHG and OPA in the upgraded system are examined.

Main results:

- Owing to the measured four times higher damage threshold, the superior nonlinear properties and a comparable pricing, LBO is preferred over DKDP as a crystal material. For the size of the LBO crystals, the largest commercially available diameter of 80 mm will be used.

- Measurements with small test crystals give an optimal thickness of 1.5 mm for the SHG crystal at the planned peak fluence of $0.19 \, \text{J/cm}^2$ (fundamental beam).
- Testing different crystal thicknesses for OPA reveals a trade-off between energy, stability and spectral smoothness. A final decision will be made based on the results of a second campaign in the near future.

6.3 Pulse Front Matching

A detailed analysis of the required pulse front matching of pump and signal beam in large non-collinear OPA systems is presented. Different schemes are discussed that introduce a pulse front tilt (PFT) in the pump beam.

Main results:

- Matching the pulse fronts of pump and signal at the second and third OPA stage is crucial to ensure spatio-temporal overlap and efficient amplification.
- A potential scheme that tilts the pump pulse front with a pair of large transmission gratings is discarded due to the low damage threshold of the tested gratings.
- Instead, an alternative scheme is suggested that introduces the required PFT by a controlled misalignment of the pump compressor. This method allows an independent control of the pump PFT at the second and third OPA stage. An additional pair of transmission gratings will adjust the tilt for the (weak) pump pulses at the first OPA stage.

References

1. T.H. Maiman, Stimulated optical radiation in ruby. Nature **187**, 493–494 (1960). https://doi.org/10.1038/187493a0
2. A. Ashkin, Acceleration and trapping of particles by radiation pressure. Phys. Rev. Lett. **24**, 156–159 (1970). https://doi.org/10.1103/PhysRevLett.24.156
3. T.W. HAnsch, Nobel lecture: Passion for precision. Rev. Mod. Phys. **78**, pp. 1297–1309 (2006). https://doi.org/10.1103/RevModPhys.78.1297
4. B.P. Abbott et al., GW150914: The advanced LIGO detectors in the era of first discoveries. Phys. Rev. Lett. **116**, 1–12 (2016). https://doi.org/10.1103/PhysRevLett.116.131103
5. A. Laubereau, W. Kaiser, Vibrational dynamics of liquids and solids investigated by picosecond light pulses. Rev. Mod. Phys. **50**, 607–665, (1978). https://doi.org/10.1103/RevModPhys.50.607
6. A.H. Zewail, Femtochemistry:atomic-scale dynamics of the chemical bond. J. Phys. Chem. A **104**, 5660–5694 (2000). https://doi.org/10.1021/jp001460h
7. A. Sanchez, R.E. Fahey, A.J. Strauss, R.L. Aggarwal, Room-temperature continuous-wave operation of a Ti:Al2O3 laser. Opt. Lett. **11**, 363–364 (1986)
8. D.E. Spence, P.N. Kean, W. Sibbett, 60-fsec pulse generation from a self-mode-locked Ti:sapphire laser. Opt. Lett. **16**, 42–44 (1991). https://doi.org/10.1364/OL.16.000042
9. U. Morgner, F.X. KArtner, S.H. Cho, Y. Chen, H.A. Haus, J. G. Fujimoto, E.P. Ippen, V. Scheuer, G. Angelow, T. Tschudi, Sub-two-cycle pulses from a Kerr-lens mode-locked Ti:sapphire laser. Opt. Lett. **24**, 411–413 (1999). https://doi.org/10.1364/OL.24.000920

10. P.F. Moulton, Spectroscopic and laser characteristics of Ti: Al2O3. J. Opt. Soc. Am. B **3**, 125–133 (1986). https://doi.org/10.1364/JOSAB.3.000125
11. Spectra-Physics, Femtopower Datasheet, http://www.spectra-physics.com/assets/client_files/files/documents/datasheets/Femtopower%20data%20sheet.pdf. Accessed 28 Mar 2017
12. Z. Gan, L. Yu, S. Li, C. Wang, X. Liang, Y. Liu, W. Li, Z. Guo, Z. Fan, X. Yuan, L. Xu, Z. Liu, Y. Xu, J. Lu, H. Lu, D. Yin, Y. Leng, R. Li, Z. Xu, 200 J high efficiency Ti : sapphire chirped pulse amplifier pumped by temporal dual- pulse. Opt. Express **25**, 5169–5178 (2017). https://doi.org/10.1364/OE.25.005169
13. P.A. Franken, A.E. Hill, C.W. Peters, G. Weinreich, Generation of optical harmonics. Phys. Rev. Lett. **7**, 118–119 (1961). https://doi.org/10.1103/PhysRevLett.7.118
14. M. Nisoli, S. DeSilvestri, O. Svelto, Generation of high energy 10 fs pulses by a new pulse compression technique. Appl. Phys. Lett. **68**, 2793–2795 (1996). https://doi.org/10.1063/1.116609
15. A.L. Cavalieri, E. Goulielmakis, B. Horvath, W. Helml, M. Schultze, M. Fiess, V. Pervak, L. Veisz, V.S. Yakovlev, M. Uiberacker, A. Apolonski, F. Krausz, R. Kienberger, Intense 1.5-cycle near infrared laser waveforms and their use for the generation of ultra-broadband soft-x-ray harmonic continua. New J. Phys. 9 (2007). https://doi.org/10.1088/1367-2630/9/7/242
16. A. McPherson, G. Gibson, H. Jara, U. Johann, T.S. Luk, I.A. McIntyre, K. Boyer, C.K. Rhodes, Studies of multiphoton production of vacuum-ultraviolet radiation in the rare gases. J. Opt. Soc. Am. B **4**, 595 (1987). https://doi.org/10.1364/JOSAB.4.000595
17. T. Brabec, F. Krausz, Intense few-cycle laser fields: Frontiers of nonlinear optics. Rev. Mod. Phys. **72**, 545–591 (2000). https://doi.org/10.1103/RevModPhys.72.545
18. E. Goulielmakis, M. Schultze, M. Hofstetter, V.S. Yakovlev, J. Gagnon, M. Uiberacker, A.L. Aquila, E.M. Gullikson, D.T. Attwood, R. Kienberger, F. Krausz, U. Kleineberg, Single-cycle nonlinear optics. Science **320**, 1614–7 (2008). https://doi.org/10.1126/science.1157846
19. A.L. Cavalieri, N. MUller, T. Uphues, V.S. Yakovlev, A. BaltuSka, B. Horvath, B. Schmidt, L. BlUmel, R. Holzwarth, S. Hendel, M. Drescher, U. Kleineberg, P.M. Echenique, R. Kienberger, F. Krausz, U. Heinzmann, Attosecond spectroscopy in condensed matter. Nature **449**, 1029–1032 (2007). https://doi.org/10.1038/nature06229
20. F. Krausz, M. Ivanov, Attosecond physics. Rev. Mod. Phys. **81**, 163–234 (2009). https://doi.org/10.1103/RevModPhys.81.163
21. A. Sommer, E.M. Bothschafter, S.A. Sato, C. Jakubeit, T. Latka, O. Razskazovskaya, H. Fattahi, M. Jobst, W. Schweinberger, V. Shirvanyan, V.S. Yakovlev, R. Kienberger, K. Yabana, N. Karpowicz, M. Schultze, F. Krausz, Attosecond nonlinear polarization and light-matter energy transfer in solids. Nature **534**, 86–90 (2016). https://doi.org/10.1038/nature17650
22. M. Bellini, C. Corsi, M.C. Gambino, Neutral depletion and beam defocusing in harmonic generation from strongly ionized media. Phys. Rev. A **64**, 1–10 (2001). https://doi.org/10.1103/PhysRevA.64.023411
23. E. Takahashi, Y. Nabekawa, T. Otsuka, M. Obara, K. Midorikawa, Generation of highly coherent submicrojoule soft x rays by high-order harmonics. Phys. Rev. A **66**, 1–4 (2002). https://doi.org/10.1103/PhysRevA.66.021802
24. G. Sansone, L. Poletto, M. Nisoli, High-energy attosecond light sources. Nat. Photonics **5**, 655–663 (2011). https://doi.org/10.1038/nphoton.2011.167
25. D.E. Rivas, M. Weidman, B. Bergues, A. Muschet, A. Guggenmos, O. Razskazovskaya, H. SchrOder, W. Helm- l, G. Marcus, R. Kienberger, U. Kleineberg, V. Pervak, P. Tzallas, D. Charalambidis, F. Krausz, L. Veisz, Generation of High-Energy Isolated Attosecond Pulses for XUV-pump/XUV-probe Experiments at 100 eV, in *High- Brightness Sources and Light-Driven Interactions 18762*, HT1B.1, (2016). https://doi.org/10.1364/HILAS.2016.HT1B.1
26. D. Umstadter, Relativistic laser-plasma interactions. J. Phys. D: Appl. Phys. **36** (2003). https://doi.org/10.1088/0022-3727/36/8/202
27. G.D. Tsakiris, K. Eidmann, J.Meyer-ter-Vehn, F. Krausz, Route to intense single attosecond pulses. New J. Phys. **8** (2006). https://doi.org/10.1088/1367-2630/8/1/019
28. U. Teubner, P. Gibbon, High-order harmonics from laser-irradiated plasma surfaces. Rev. Mod. Phys. **81**, 445–479 (2009). https://doi.org/10.1103/RevModPhys.81.445

29. C. Thaury, F. QuErE, High-order harmonic and attosecond pulse generation on plasma mirrors: basic mechanisms. J. Phys. B: At. Mol. Opt. Phys. **43**, 213001 (2010). https://doi.org/10.1088/0953-4075/43/21/213001

30. P. Heissler, R. HOrlein, J.M. Mikhailova, L. Waldecker, P. Tzallas, A. Buck, K. Schmid, C.M.S. Sears, F. Krausz, L. Veisz, M. Zepf, G.D. Tsakiris, Few-cycle driven relativistically oscillating plasma mirrors: A source of intense isolated attosecond pulses. Phys. Rev. Lett. **108**, 235003 (2012). https://doi.org/10.1103/PhysRevLett.108.235003

31. E.W. Gaul, M. Martinez, J. Blakeney, A. Jochmann, M. Ringuette, D. Hammond, T. Borger, R. Escamilla, S. Douglas, W. Henderson, G. Dyer, A. Erlandson, R. Cross, J. Caird, C. Ebbers, T. Ditmire, Demonstration of a 1.1 petawatt laser based on a hybrid optical parametric chirped pulse amplification/mixed Nd:glass amplifier. Appl. Opt. **49**, 1676–1681 (2010). https://doi.org/10.1364/AO.49.001676

32. W.P. Leemans, J. Daniels, A. Deshmukh, A.J. Gonsalves, A. Magana, H.-S. Mao, D.E. Mittelberger, K. Naka-Mura, J. R. Riley, D. Syversrud, C. TOth, N. Ybarrolaza, BELLA laser and operations, in *Proceedings of PAC* (2013), pp. 1097–1100

33. F. Böhle, M. Kretschmar, A. Jullien, M. Kovacs, M. Miranda, R. Romero, H. Crespo, U. Morgner, P. Simon, R. Lopez-Martens, T. Nagy, Compression of CEP-stable multi-mJ laser pulses down to 4 fs in long hollow fibers. Laser Phys. Lett. **11**, 095401 (2014). https://doi.org/10.1088/1612-2011/11/9/095401

34. O. Hort, A. Dubrouil, A. Cabasse, S. Petit, E. Mével, D. Descamps, E. Constant, Postcompression of high-energy terawatt-level femtosecond pulses and application to high-order harmonic generation. J. Opt. Soc. Am. B **32**D, 1055 (2015). https://doi.org/10.1364/JOSAB.32.001055

35. S. Mironov, E. Khazanov, G. Mourou, Pulse shortening and ICR enhancement for PW-class lasers, in *Specialty Optical Fibers*, JTu3A.24 (2014)

36. G. Mourou, S. Mironov, E. Khazanov, A. Sergeev, Single cycle thin film compressor opening the door to Zeptosecond-Exawatt physics. Eur. Phys. J. Spec. Top. **223**, 1181–1188 (2014). https://doi.org/10.1140/epjst/e2014-02171-5

37. C. Hooker, Y. Tang, O. Chekhlov, J. Collier, E. Divall, K. Ertel, S. Hawkes, B. Parry, P.P. Rajeev, Improving coherent contrast of petawatt laser pulses. Opt. Express **19**, 2193–2203 (2011). https://doi.org/10.1364/OE.19.002193

38. L. Yu, Z. Xu, X. Liang, L. Xu, W. Li, C. Peng, Z. Hu, C. Wang, X. Lu, Y. Chu, Z. Gan, X. Liu, Y. Liu, X. Wang, H. Lu, D. Yin, Y. Leng, R. Li, Z. Xu, Optimization for high-energy and high-efficiency broadband optical parametric chirped-pulse amplification inLBOnear 800 nm. Opt. Lett. **40**, 3412 (2015). https://doi.org/10.1364/OL.40.003412

39. F. Lureau, S. Laux, O. Casagrande, O. Chalus, A. Pellegrina, G. Matras, C. Radier, G. Rey, S. Ricaud, S. Herriot, P. Jougla, M. Charbonneau, P. Duvochelle, C. Simon-Boisson, Latest results of 10 petawatt laser beamline for ELI nuclear physics infrastructure, in *Proceedings of the SPIE*, vol. 9726 (2016). https://doi.org/10.1117/12.2213067

40. D.N. Papadopoulos, J. Zou, C.L. Blanc, G. Ch, A. Beluze, N. Lebas, P. Monot, F. Mathieu, P. Audebert, The Apollon 10PWlaser: experimental and theoretical investigation of the temporal characteristics. High Power Laser Sci. Eng. **4**, 1–7 (2016). https://doi.org/10.1017/hpl.2016.34

41. J.A. Armstrong, N. Bloembergen, J. Ducuing, P.S. Pershan, Interactions between light waves in a nonlinear dielectric. Phys. Rev. **127**, 1918–1939 (1962). https://doi.org/10.1103/PhysRev.127.1918

42. G. Cerullo, S. De Silvestri, Ultrafast optical parametric amplifiers. Rev. Sci. Instrum. **74**, 1–18 (2003). https://doi.org/10.1063/1.1523642

43. O.V. Chekhlov, J.L. Collier, I.N. Ross, P.K. Bates, M. Notley, C. Hernandez-Gomez, W. Shaikh, C. N. Dan- son, D. Neely, P. Matousek, S. Hancock, L. Cardoso, 35 J broadband femtosecond optical parametric chirped pulse amplification system. Opt. Lett. **31**, 3665 (2006). https://doi.org/10.1364/OL.31.003665

44. L. Veisz, D. Rivas, G. Marcus, X. Gu, D. Cardenas, J. Xu, J. Mikhailova, A. Buck, T. Wittmann, C.M.S. Sears, D. Herrmann, O. Razskazovskaya, V. Pervak, F. Krausz, Multi-10-TWsub-5-fs optical parametric synthesizer, in *2014 IEEE Photonics Conference* vol. 163, (2014), pp. 510–511. https://doi.org/10.1109/IPCon.2014.6995473

45. S. Karsch, Z. Major, J. FUlOp, I. Ahmad, T.-J.Wang, A. Henig, S. Kruber, R. Weingartner, M. Siebold, J. Hein, C. Wandt, S. Klingebiel, J. Osterhoff, R. HOrlein, F. Krausz, The petawatt field synthesizer: a new approach to ultrahigh field generation. Adv. Sol.-State Photonics WF1 (2008). https://doi.org/10.1364/ASSP.2008.WF1

46. Z. Major, S.A. Trushin, I. Ahmad, M. Siebold, C. Wandt, S. Klingebiel, T.-J. Wang, J.A. FUlOp, A. Henig, S. Kruber, R. Weingartner, A. Popp, J. Osterhoff, R. HOrlein, J. Hein, V. Pervak, A. Apolonski, F. Krausz, S. Karsch, Basic concepts and current status of the petawatt field synthesizer-a new approach to ultrahigh field generation. Rev. Laser Eng. **37**, 431–436 (2009). https://doi.org/10.2184/lsj.37.431

47. B.C. Stuart, M.D. Feit, A.M. Rubenchik, B.W. Shore, M.D. Perry, Laser-induced damage in dielectrics with nanosecond to subpicosecond pulses. Phys. Rev. Lett. **74**, 2248–2251 (1995). https://doi.org/10.1103/PhysRevLett.74.2248

48. C. Skrobol, I. Ahmad, S. Klingebiel, C. Wandt, S.A. Trushin, Z. Major, F. Krausz, S. Karsch, Broadband amplification by picosecond OPCPA in DKDP pumped at 515 nm. Opt. Express **20**, 4619–4629 (2012)

49. C. Skrobol, *High-Intensity, Picosecond-Pumped, Few-CycleOPCPA* (Ludwig-Maximilians-Universität München, PhDthesis, 2014)

50. L. Veisz, D. Rivas, G. Marcus, X. Gu, D. Cardenas, J. Mikhailova, A. Buck, T. Wittmann, C.M.S. Sears, S.W. Chou, J. Xu, G. Ma, D. Herrmann, O. Razskazovskaya, V. Pervak, F. Krausz, Generation and applications of sub-5-fs multi-10-TW light pulses, in *Pacific Rim Conference on Lasers and Electro-Optics, CLEO-Technical Digest* (2013). https://doi.org/10.1109/CLEOPR.2013.6600068

51. J. Moses, C. Manzoni, S.-W. Huang, G. Cerullo, F.X. Kaertner, Temporal optimization of ultrabroadband high-energy OPCPA. Opt. Express **17**, 5540 (2009). https://doi.org/10.1364/OE.17.005540

52. D. Strickland, G. Mourou, Compression of amplified chirped optical pulses. Opt. Commun. **56**, 219–221 (1985). https://doi.org/10.1016/0030-4018(85)90120-8

53. I. Ahmad, S.A. Trushin, Z. Major, C. Wandt, S. Klingebiel, T.J. Wang, V. Pervak, A. Popp, M. Siebold, F. Krausz, S. Karsch, Frontend light source for short-pulse pumped OPCPA system. Appl. Phys. B Lasers Opt. **97**, 529–536 (2009). https://doi.org/10.1007/s00340-009-3599-4

54. I. Ahmad, Development of an optically synchronized seed source for a high-power few-cycle OPCPA system, PhD thesis, Ludwig-Maximilians-Universität München, 2011

55. S. Klingebiel, C. Wandt, C. Skrobol, I. Ahmad, S.A. Trushin, Z. Major, F. Krausz, S. Karsch, High energy picosecond Yb:YAG CPA system at 10 Hz repetition rate for pumping optical parametric amplifiers. Opt. Express **19**, 421–427 (2011)

56. S. Klingebiel, I. Ahmad, C. Wandt, C. Skrobol, S.A. Trushin, Z. Major, F. Krausz, S. Karsch, Experimental and theoretical investigation of timing jitter inside a stretcher-compressor setup. Opt. Express **20**, 3443–3455 (2012). https://doi.org/10.1364/OE.20.003443

57. S. Klingebiel, Picosecond PumpDispersionManagement and Jitter Stabilization in a Petawatt-Scale Few-Cycle OPCPA System, PhD thesis, Ludwig-Maximilians-Universität München, 2013

58. C. Wandt, S. Klingebiel, S. Keppler, M. Hornung, C. Skrobol, A. Kessel, S. a. Trushin, Z. Major, J. Hein, M. C. Kaluza, F. Krausz, S. Karsch, Development of a Joule-class Yb:YAG amplifier and its implementation in a CPA system generating 1 TW pulses. Laser Photonic Rev. **881**, 875–881 (2014). https://doi.org/10.1002/lpor.201400040

59. C. Wandt, Development of a Joule-class Yb:YAG amplifier and its implementation in a CPA system generating 1 TWpulses, PhD thesis, Ludwig-Maximilians-Universität München, 2014

Chapter 2
Fundamentals

This thesis is dedicated to the generation and amplification of intense, ultrashort light pulses. For a thorough understanding of the physical effects encountered when working with such pulses, the following chapter provides a mathematical description of the propagation and interaction of light. Furthermore, some experimental tools for the temporal characterization of ultrashort pulses are introduced.

2.1 Mathematical Description of Ultrashort Light Pulses

On the next pages we will derive the first-order propagation equation of electromagnetic fields and discuss its spatial and temporal solutions, i.e. the mathematical approximations to experimentally observable light pulses.

The description and nomenclature follows and combines the well-known books written by Boyd [1], Sutherland [2] and Trebino [3] as well as the notes of a lecture on computational photonics held by Karpowicz [4].

2.1.1 Derivation of the First-Order Propagation Equation

Since optical pulses are a special form of electromagnetic fields, their description starts with Maxwell's equations. Typically, pulses propagate in regions containing no free charges ($\rho = 0$) or free currents ($j = 0$) and hence we can limit ourselves to this case:

$$\nabla \cdot \boldsymbol{D} = 0 \qquad\qquad \nabla \cdot \boldsymbol{B} = 0$$

$$\nabla \times \boldsymbol{E} = -\partial_t \boldsymbol{B} \qquad\qquad \nabla \times \boldsymbol{H} = \partial_t \boldsymbol{D} \qquad (2.1)$$

© Springer International Publishing AG, part of Springer Nature 2018
A. Kessel, *Generation and Parametric Amplification of Few-Cycle Light Pulses at Relativistic Intensities*, Springer Theses,
https://doi.org/10.1007/978-3-319-92843-2_2

where $E = E(r, t)$ is the space- and time-dependent electric field, $B(r, t)$ is the magnetic, $H(r, t)$ the magnetizing and $D(r, t)$ the displacement field. For propagation in nonmagnetic media it is:

$$B = \mu_0 H. \tag{2.2}$$

The displacement can be expanded to

$$D = \varepsilon_0 E + P \tag{2.3}$$

with the polarization P that is defined as

$$
\begin{aligned}
P &= \varepsilon_0 \left(\chi^{(1)} E + \chi^{(2)} EE + \chi^{(3)} EEE + \cdots \right) \\
&= P^{(1)} + P^{(NL)}
\end{aligned} \tag{2.4}
$$

where $\chi^{(n)}$ is the n-th order susceptibility (tensor) of the medium and $P^{(1)}$ and $P^{(NL)}$ are the linear and nonlinear polarization. Combining Eqs. (2.1)–(2.3) yields the well-known general wave equation:

$$\nabla \times \nabla \times E + \frac{1}{c^2} \partial_t^2 E = -\frac{1}{\varepsilon_0 c^2} \partial_t^2 P \tag{2.5}$$

It is common to use the approximation

$$\nabla \times \nabla \times E = \nabla(\nabla E) - \Delta E \approx -\Delta E \tag{2.6}$$

which is valid if ∇E is zero or very small. In linear optics and for homogeneous media this is fulfilled as we have $P = P^{(1)} = \varepsilon_0 \chi^{(1)} E$ and therefore the Eqs. (2.1) and (2.3) yield $\nabla E \propto \nabla D = 0$. In nonlinear optics $P \not\propto E$ and thus $\nabla E \not\propto \nabla D$. However, for a transverse infinite plane wave it holds that $\nabla E = 0$. Hence, for a field close to a transverse, infinite plane wave (i.e. a collimated or loosely focused laser beam) ∇E is close to zero.

Applying now the approximation of Eq. (2.6) on Eq. (2.5) one obtains:

$$\Delta E - \frac{1}{c^2} \partial_t^2 E = \frac{1}{\varepsilon_0 c^2} \partial_t^2 P. \tag{2.7}$$

Fourier transforming this equation and applying $\partial_t \to -i\omega$ yields

$$\Delta \tilde{E} + \frac{\omega^2}{c^2} \tilde{E} = -\frac{\omega^2}{\varepsilon_0 c^2} \tilde{P} \tag{2.8}$$

where the tilde-notation indicates fields in the frequency domain, e.g.:

$$\tilde{E} = \tilde{E}(r,\omega) = \underset{t\to\omega}{\mathfrak{Fou}}\{E(r,t)\} = \int\limits_{-\infty}^{\infty} E(r,t)\,e^{-i\omega t}\,dt. \tag{2.9}$$

Splitting the polarization in its linear and nonlinear parts we can rewrite Eq. (2.8) as

$$\Delta\tilde{E} + \frac{\omega^2}{c^2}\tilde{E} = -\frac{\omega^2}{\varepsilon_0 c^2}\left(\tilde{P}^{(1)} + \tilde{P}^{NL}\right)$$

$$\Leftrightarrow \qquad \Delta\tilde{E} + \frac{\omega^2}{c^2}\left(\tilde{E} + \tilde{P}^{(1)}/\varepsilon_0\right) = -\frac{\omega^2}{\varepsilon_0 c^2}\tilde{P}^{NL}$$

$$\Leftrightarrow \qquad \Delta\tilde{E} + \frac{\omega^2}{c^2}\left(1 + \chi^{(1)}(\omega)\right)\tilde{E} = -\frac{\omega^2}{\varepsilon_0 c^2}\tilde{P}^{NL}$$

$$\Leftrightarrow \qquad \Delta\tilde{E} + \frac{\omega^2}{c^2}\varepsilon^{(1)}(\omega)\tilde{E} = -\frac{\omega^2}{\varepsilon_0 c^2}\tilde{P}^{NL}$$

$$\Leftrightarrow \qquad \Delta\tilde{E} + k^2(\omega)\tilde{E} = -\frac{\omega^2}{\varepsilon_0 c^2}\tilde{P}^{NL} \tag{2.10}$$

with the permittivity of the medium $\varepsilon^{(1)}$ and the **wavenumber** $k = \frac{\omega}{c}\sqrt{\varepsilon^{(1)}}$. Assuming that the wave propagates in z-direction, one can separate the Laplace operator into axial and transversal derivatives:

$$\left[\partial_z^2 + \Delta_\perp + k^2(\omega)\right]\tilde{E} = -\frac{\omega^2}{\varepsilon_0 c^2}\tilde{P}^{NL} \tag{2.11}$$

In this partial differential equation (PDE) the important terms are already disentangled but it still contains a second-order derivative of z which makes it computationally disadvantageous. To transform it into a first-order PDE, one can use the ansatz suggested by Brabec and Krausz [5]

$$\tilde{E}(r,\omega) = e^{ik(\omega)z}\tilde{U}(r,\omega) \tag{2.12}$$

i.e. the electric field is split into a slowly and a rapidly varying function with z. Substituting $\tilde{E}$ in Eq. (2.11) and performing the differentiation one obtains:

$$\left[\partial_z^2 + 2ik\,\partial_z + \Delta_\perp\right]\tilde{U} = -e^{-ikz}\frac{\omega^2}{\varepsilon_0 c^2}\tilde{P}^{NL} \tag{2.13}$$

In order to neglect the second-order derivative in this equation the condition

$$\left|\partial_z^2\,\tilde{U}\right| \ll 2k\left|\partial_z\,\tilde{U}\right| \tag{2.14}$$

must be fulfilled. This approximation is called "slowly varying envelope approximation" (SVEA) and requires that nonlinear and diffraction effects are weak, i.e. that the envelope of the pulse is changing slowly[1] during propagation. Furthermore, the SVEA in general applies only if the pulse duration is long compared to one oscillation of the electric field. It has been shown in [6], however, that under the assumption of weak[2] dispersion far from resonances the approximation in Eq. (2.14) can also be used for the propagation of few-cycle pulses. This is known as the "slowly evolving wave approximation" (SEWA).

Hence, Eq. (2.13) can be approximated to:

$$\left[2ik\, \partial_z + \Delta_\perp\right] \tilde{U} = -e^{-ikz}\frac{\omega^2}{\varepsilon_0 c^2}\tilde{P}^{\mathrm{NL}} \qquad (2.15)$$

Re-substituting $\tilde{U}$ by $\tilde{E}$ yields

$$\left[2ik\, \partial_z + \Delta_\perp\right] e^{-ikz}\tilde{E} = -e^{-ikz}\frac{\omega^2}{\varepsilon_0 c^2}\tilde{P}^{\mathrm{NL}}$$

$$\Leftrightarrow \quad 2k^2 e^{-ikz}\tilde{E} + 2ike^{-ikz}\partial_z\tilde{E} + e^{-ikz}\Delta_\perp\tilde{E} = -e^{-ikz}\frac{\omega^2}{\varepsilon_0 c^2}\tilde{P}^{\mathrm{NL}} \qquad (2.16)$$

and finally results in the *first-order propagation equation (FOPE)*:

$$\partial_z \tilde{E}(r, \omega) = \underbrace{ik(\omega)\; \tilde{E}(r, \omega)}_{\substack{\text{dispersion,}\\\text{absorption}}} + \underbrace{\frac{i}{2k(\omega)}\Delta_\perp \tilde{E}(r, \omega)}_{\text{diffraction}} + \underbrace{\frac{i\omega^2}{2\varepsilon_0 c^2 k(\omega)}\tilde{P}^{\mathrm{NL}}(r, \omega)}_{\text{nonlinear effects}}$$

$$(2.17)$$

As denoted below the individual terms, all important spatial and temporal effects of pulse propagation are present in this equation and nicely disentangled. The physical meaning of these terms will be described in the next sections in more detail.

Summary

Let us briefly summarize the approximations and assumptions that lead to Eq. (2.17) in order to clarify under which conditions it is valid:

- The medium of light propagation contains *no free charges or currents* and is *non-magnetic*.
- The propagation takes place in a *homogeneous medium*. In particular there are no abrupt changes of physical properties in the volume of interest (such as on interfaces).

---

[1]More precisely: there is only a small change of $\tilde{U}$ within a distance of one wavelength $\lambda = 2\pi/k$.

[2]The mathematical requirement is that the difference between phase velocity $v_{\mathrm{p}} = \frac{\omega}{k}$ and group velocity $v_{\mathrm{g}} = \frac{\partial \omega}{\partial k}$ is small compared to the latter.

- There are *no counter-propagating beams* (e.g. reflections from surfaces) that interact/interfere with the beam of interest. This constraint is a consequence of the PDE-type of Eq. (2.17): it restricts itself to a propagating of the field—known at a certain z-position—in forward z-direction in contrast to the general wave equation Eq. (2.5) that calculates the field at all points in space and time as a whole. However, dealing with just one z-slice at a time makes the first-order propagation equation computationally very advantageous.
- The *paraxial* approximation applies, i.e. the simulated beams are neither strongly focused nor intersect at a large angle.
- During propagation the variation of the field amplitude due to nonlinear effects and the variation of the temporal shape of the pulse due to dispersion is slow (*SEWA*).

Despite this apparently large number of restrictions the FOPE applies for many practical cases in the lab such as the propagation and nonlinear interaction of collimated or loosely focused laser pulses in homogeneous media (vacuum, gases, solids). It therefore serves as the basis for the following theoretical considerations and will also be used for numerical simulations.

2.1.2 Spatial Description and Diffraction of Light Pulses

Neglecting for the moment the dispersion term and nonlinear effects in Eq. (2.17), one obtains the paraxial Helmholtz equation that approximates the spatial propagation of either weak and monochromatic light or light that propagates in vacuum:

$$\Delta_\perp \tilde{E}(r, \omega) + 2ik \, \partial_z \tilde{E}(r, \omega) = 0 \tag{2.18}$$

The most prominent solution to this equation is the monochromatic *Gaussian beam* that—assuming linear polarization in x-direction—is described by

$$\tilde{E}(r_\perp, z) = E_0 \, \hat{x} \, \frac{w_0}{w(z)} \, e^{-r_\perp^2/w(z)^2} \, e^{-ikr_\perp^2/(2R(z))} \, e^{i(\zeta(z) - kz)} \tag{2.19}$$

where $r_\perp = \sqrt{x^2 + y^2}$, E_0 is the electric field amplitude, w_0 is the waist size and $w(z)$ is the position-dependent beam size:

$$w(z) = w_0 \sqrt{1 + \left(\frac{z}{z_R}\right)^2} \tag{2.20}$$

with the Rayleigh length $z_R = \frac{\pi w_0}{\lambda}$, where $\lambda = \frac{2\pi c}{\omega}$ is the wavelength. The radius of curvature of the wavefront $R(z)$ evolves as

$$R(z) = z \left(1 + \left(\frac{z_R}{z}\right)^2\right) \tag{2.21}$$

and the Gouy phase is

$$\zeta(z) = \arctan\left(\frac{z}{z_R}\right). \tag{2.22}$$

Starting from the position of the waist ($z = 0$) the beam size w increases according to Eq. (2.20) because of diffraction. The *shape* of the transversal profile of a Gaussian beam, however, does not change during propagation. This is reflected by the expression for the intensity

$$I(r_\perp, z) = I_0 \left(\frac{w_0}{w(z)}\right)^2 e^{-2r_\perp^2/w(z)^2} \propto \left|\tilde{E}(r_\perp, z)\right|^2. \tag{2.23}$$

where the functional dependence on the transverse coordinate $r_\perp$ remains unchanged for all positions z. This invariance of shape during propagation that is depicted in Fig. 2.1a makes Gaussian beams useful for many applications and simulations. On the other hand, for nonlinear conversion processes that will be discussed later in this chapter, it is often advantageous to have a constant intensity over a large area of the beam profile. The extreme case would be a *flat-top beam* but strong diffraction effects (see Fig. 2.1b) due to the steep edges complicate beam transport and limit its practical usability. A compromise between Gaussian and flat-top beam is the *nth-order super-Gaussian* with an intensity distribution of

$$I_n(r_\perp, 0) = I_0 e^{-r_\perp^n/2\sigma^n} \tag{2.24}$$

where σ is a beam size parameter. The rather smooth propagation and the intensity plateau at the center of the beam as shown for a 6th-order super-Gaussian in Fig. 2.1c make the super-Gaussian a frequently used beam shape for nonlinear applications.

The size of laser beams is commonly specified in terms of the *full width at half maximum (FWHM)* or the *$1/e^2$-diameter*. Another useful definition is the *D86 width* that gives the diameter of an aperture transmitting 86% of the total beam power. For conversion between these quantities see Table 2.1. In order to estimate the encircled power for arbitrary aperture diameters, Fig. 2.2 can be used.

Often one is interested in the peak intensity of a laser pulse with a given energy E, beam diameter D_{FWHM} and pulse duration τ_{FWHM}. Assuming a Gaussian envelope in time, the peak intensity in the near field for the different beam shapes can be calculated as

$$I_{max} = \kappa \frac{E}{D_{FWHM}^2 \, \tau_{FWHM}} \tag{2.25}$$

where the factor κ depends on the shape (see again Table 2.1).

For laser-plasma interactions, an important parameter is the **normalized vector potential** of a laser pulse with central wavelength λ:

$$a_0 = \sqrt{\frac{I \, [\text{W/cm}^2]}{1.37 \times 10^{18}}} \cdot \lambda \, [\mu\text{m}] \, . \tag{2.26}$$

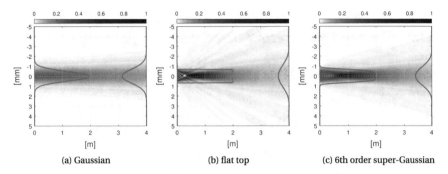

(a) Gaussian   (b) flat top   (c) 6th order super-Gaussian

Fig. 2.1 Diffraction effects for different beam profiles. Direction of propagation is along the x-axis, lateral size of the beam is along the y-axis. Propagation is simulated in 3D with monochromatic beams ($\lambda = 800$ nm). Plotted lines show the near-field and far-field intensity distribution (cross section through the center of the beam). Colors indicate the intensity distribution during propagation

Table 2.1 Conversion table for different beam size parameters/definitions

	σ (Eq. (2.24))	$1/e^2$-diameter	D86 width	κ (Eq. (2.25))
2nd order Gauss	$0.42 \times D_{\text{FWHM}}$	$1.7 \times D_{\text{FWHM}}$	$1.7 \times D_{\text{FWHM}}$	$\left(\frac{4\ln 2}{\pi}\right)^{3/2} \approx 0.83$
4th order Gauss	$0.46 \times D_{\text{FWHM}}$	$1.3 \times D_{\text{FWHM}}$	$1.1 \times D_{\text{FWHM}}$	1.12
6th order Gauss	$0.47 \times D_{\text{FWHM}}$	$1.2 \times D_{\text{FWHM}}$	$1.0 \times D_{\text{FWHM}}$	1.19
Flat top	$-$	$1 \times D_{\text{FWHM}}$	$0.93 \times D_{\text{FWHM}}$	$\left(\frac{64\ln 2}{\pi^3}\right)^{1/2} \approx 1.2$

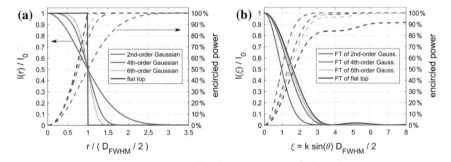

Fig. 2.2 **a** Intensity and encircled power for different beam shapes. Solid lines show the intensity cross sections, dashed lines the encircled power. **b** Same plot as in (**a**) but with the respective 2D-Fourier-transformed beam shapes. For the x-axis scaling we used the FWHM diameter D_{FWHM} of the original beam, the wavenumber k and the angle of observation ϑ

Its amplitude indicates whether the momentum gain of an electron within one field cycle of this laser pulse exceeds the electron rest mass: For $a_0 \gg 1$ the electron motion is considered as relativistic for $a_0 \ll 1$ it is non-relativistic.

2.1.3 Temporal and Spectral Description of Light Pulses

The time dependent electric field of a laser pulse at a fixed point in space can be expressed as

$$E(t) = \mathcal{E}(t)\ \cos\left(\Phi(t)\right) \tag{2.27}$$

$$= \frac{\mathcal{E}(t)}{2}\left(e^{-i\Phi(t)} + e^{i\Phi(t)}\right) \tag{2.28}$$

with the *field envelope* $\mathcal{E}(t)$ and the *temporal phase* $\Phi(t)$. The *intensity* $I(t)$ of this field is by definition:

$$I(t) = \frac{c\,\varepsilon_0}{2}\mathcal{E}^2(t) \tag{2.29}$$

As an example Fig. 2.3 shows two pulses with a Gaussian envelope and a linear temporal phase $\Phi(t) = \omega_0 t + \Phi_0$, where the frequency ω_0 is called *carrier frequency* and the relative offset Φ_0 between envelope and carrier wave is the *carrier-envelope-phase (CEP)*. While for pulses with many electric field oscillations (Fig. 2.3a) the CEP has little influence on the amplitude of each half-cycle the effect is significant for few-cycle pulses (Fig. 2.3b). Hence, the measurement and stabilization of the CEP plays an important role when performing experiments with ultrashort pulses [7].

For mathematical convenience the conjugate term in Eq. (2.28) is generally omitted and only the complex field $E^+(t)$ is further used:

$$E^+(t) = \mathcal{E}(t)\,e^{-i\Phi(t)} \tag{2.30}$$

$$= \sqrt{\frac{2\,I(t)}{c\,\varepsilon_0}}\,e^{-i\Phi(t)} \tag{2.31}$$

The real electric field, however, can always be retrieved by adding the complex conjugate:

$$E(t) = \frac{1}{2}\left(E^+(t) + E^-(t)\right) = \frac{1}{2}\left(E^+(t) + \text{c.c.}\right) \tag{2.32}$$

The Fourier transformation of $E^+(t)$ yields the electric field in the frequency domain:

$$\tilde{E}^+(\omega) = \underset{t \to \omega}{\mathfrak{Fou}} \left\{ E^+(t) \right\}$$

$$= \int\limits_{-\infty}^{\infty} E^+(t)\, e^{-i\omega t}\, dt$$

$$=: \sqrt{S(\omega)}\, e^{-i\varphi(\omega)} \tag{2.33}$$

where $S(\omega)$ is the *spectral intensity* ("spectrum") and $\varphi(\omega)$ is the *spectral phase*. Being the Fourier transform of each other, both $E^+(t)$ and $\tilde{E}^+(\omega)$ fully characterize the laser pulse. In accordance with Eqs. (2.31) and (2.33) the intensity and phase in the respective domain can be calculated as:

$$I(t) = \frac{c\,\varepsilon_0}{2}\left|E^+(t)\right|^2 \qquad\qquad S(\omega) = \left|\tilde{E}^+(\omega)\right|^2$$

$$\Phi(t) = -\arctan\left(\frac{\mathrm{Im}\left\{E^+(t)\right\}}{\mathrm{Re}\left\{E^+(t)\right\}}\right) \qquad \varphi(\omega) = -\arctan\left(\frac{\mathrm{Im}\left\{\tilde{E}^+(\omega)\right\}}{\mathrm{Re}\left\{\tilde{E}^+(\omega)\right\}}\right) \tag{2.34}$$

Note that in dielectric media the intensity is given by $I(t) = \frac{c\,n\,\varepsilon_0}{2}\left|E^+(t)\right|^2$ with the refractive index n that will be discussed in the the next section.

From the temporal phase one can derive another useful quantity, the *instantaneous frequency* $\omega_{\mathrm{inst}}(t)$:

$$\omega_{\mathrm{inst}}(t) = -\frac{d\Phi(t)}{dt} \tag{2.35}$$

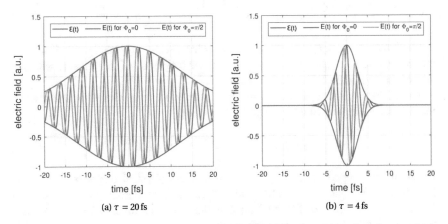

(a) $\tau = 20\,\mathrm{fs}$ (b) $\tau = 4\,\mathrm{fs}$

Fig. 2.3 Temporal waveform of optical pulses with different pulse durations τ featuring **a** multiple or **b** few oscillations of the electric field. Curves for two values of the CEP Φ_0 are shown. The carrier frequency of all pulses is $\omega_0 = 2.35\,\mathrm{PHz}$ ($\lambda_0 = 800\,\mathrm{nm}$)

It specifies the oscillation frequency of the electric field at a certain point in time and illustrates the meaning of the temporal phase.

While in this section only scalar fields $E(t)$ have been used, the generalization to a vectorial description for linearly polarized pulses is simple as the field $\mathbf{E}(t)$ can be decomposed into its—again scalar—polarization components $E_x(t)$, $E_y(t)$ and $E_z(t)$. For the treatment of elliptically polarized fields the Jones calculus can be used as described for example in [8].

2.1.4 Dispersion and Absorption

Dispersion is a physical effect that—as the name suggests—leads to a separation of spectral light components in time and space. As its impact increases with the spectral bandwidth, the effect constitutes a major challenge in ultra-fast optics where pulse spectra may span tens or hundreds of nanometers.

The term "dispersion" is often used in a broader sense to indicate any dependence of light propagation on the wavelength. Strictly speaking, however, it is defined as the wavelength/frequency-dependence of the **phase velocity** of light

$$v_p = v_p(\omega) = \frac{\omega}{k(\omega)} = \frac{\omega}{n(\omega)k_{vac}} = \frac{\omega}{n(\omega)\omega/c} = \frac{c}{n(\omega)} \tag{2.36}$$

where $n(\omega)$ is the **refractive index** and c is the vacuum speed of light. A related quantity of equal importance for the propagation of optical pulses in dispersive media is the **group velocity**

$$v_{gr}(\omega) = \frac{\partial \omega}{\partial k(\omega)} = \left(\frac{\partial k(\omega)}{\partial \omega}\right)^{-1} = \frac{c}{n(\omega) + \omega \frac{\partial n(\omega)}{\partial \omega}} \tag{2.37}$$

that indicates the propagation speed of the pulse envelope. Notably, the difference between phase and group velocity of the carrier frequency of the pulse results in a shift of the absolute CEP during propagation.

The refractive index itself is a material property and is related to the relative permittivity ε_r and the linear susceptibility $\chi^{(1)}$ by:

$$n(\omega) = \sqrt{\varepsilon_r(\omega)} = \sqrt{1 + \chi^{(1)}(\omega)} \tag{2.38}$$

The frequency-dependence of $\chi^{(1)}$, ε_r and n is the physical origin of dispersion. This dependence can be well approximated for most materials with the so-called "Sellmeier equations". In Table 2.2 the equations and coefficients for a few typical media are listed that were used in this work.

In general the refractive index is a complex number $n_{cplx} = n + i\kappa$ with the imaginary part $\kappa(\omega)$ that defines the **absorption** of light that propagates in a material.

Table 2.2 Sellmeier equations and coefficients of selected materials to calculate the wavelength-dependent refractive index. By convention the unit of the wavelength λ is μm

Material	Refractive index	
air	$n = 1 + \frac{0.0579}{238.0185 - 1/\lambda^2} + \frac{0.00167917}{57.362 - 1/\lambda^2}$	[9]
BaF$_2$	$n = \sqrt{1 + 0.33973 + \frac{0.81070\,\lambda^2}{\lambda^2 - 0.10065^2} + \frac{0.19652\,\lambda^2}{\lambda^2 - 29.87^2} + \frac{4.52469\,\lambda^2}{\lambda^2 - 53.82^2}}$	[10]
BBO	$n_e = \sqrt{2.3753 + \frac{0.01224}{\lambda^2 - 0.01667} - 0.01627\,\lambda^2 + 0.0005716\,\lambda^4 - 0.00006305\,\lambda^6}$	[11]
	$n_o = \sqrt{2.7359 + \frac{0.01878}{\lambda^2 - 0.01822} - 0.01471\lambda^2 + 0.0006081\,\lambda^4 - 0.00006740\,\lambda^6}$	
CaF$_2$	$n =$	[12]
	$\sqrt{1 + \frac{0.443749998\,\lambda^2}{\lambda^2 - 0.00178027854} + \frac{0.444930066\,\lambda^2}{\lambda^2 - 0.00788536061} + \frac{0.150133991\,\lambda^2}{\lambda^2 - 0.0124119491} + \frac{8.85319946\,\lambda^2}{\lambda^2 - 2752.28175}}$	
FS	$n = \sqrt{1 + \frac{0.6961663\,\lambda^2}{\lambda^2 - 0.0684043^2} + \frac{0.4079426\,\lambda^2}{\lambda^2 - 0.1162414^2} + \frac{0.8974794\,\lambda^2}{\lambda^2 - 9.896161^2}}$	[13]
LBO	$n_x = \sqrt{2.4542 + \frac{0.01125}{\lambda^2 - 0.01135} - 0.01388\,\lambda^2}$	[14]
	$n_y =$	
	$\sqrt{2.5390 + \frac{0.01277}{\lambda^2 - 0.01189} - 0.01848\,\lambda^2 + 0.000043025\,\lambda^4 - 0.000029131\,\lambda^6}$	
	$n_z =$	
	$\sqrt{2.5865 + \frac{0.01310}{\lambda^2 - 0.01223} - 0.01861\,\lambda^2 + 0.000045778e\,\lambda^4 - 0.000032526\,\lambda^6}$	

Together with dispersion it enters the FOPE via the frequency-dependent wavenumber $k(\omega) = k_{vac}\, n_{cplx}(\omega)$. During propagation the electric field strength decreases due to absorption, where the penetration depth L_α is the distance after which the intensity $(I \sim |E|^2)$ has dropped to $1/e$ of its initial value. It is defined by $L_\alpha = 1/\alpha$ with the absorption coefficient α:

$$\alpha(\omega) = \frac{4\pi\,\kappa(\omega)}{\lambda} \tag{2.39}$$

Tabulated values for α can be found in the literature, e.g. [15, 16].

The contribution from dispersion to the FOPE leads to an accumulated spectral phase $\varphi(\omega)$:

$$\varphi(\omega) = n(\omega)\,\frac{\omega}{c}\,L \tag{2.40}$$

where L is the propagation distance in a medium with the refractive index $n(\omega)$. It is common to write $\varphi(\omega)$ as a Taylor expansion around a central frequency ω_0:

$$\varphi(\omega) = \sum_{m=0}^{\infty} \frac{(\omega - \omega_0)^m}{m!} \left(\partial_\omega^m \varphi(\omega) \right)_{\omega = \omega_0} = \sum_{m=0}^{\infty} \frac{(\omega - \omega_0)^m}{m!}\, D_m(\omega_0) \tag{2.41}$$

The dispersion coefficients D_m in this expansion are named as follows:

$$D_0 = \varphi(\omega)\Big|_{\omega=\omega_0} \qquad = \text{zeroth-order phase} \qquad = \varphi(\omega_0)$$

$$D_1 = \partial_\omega \varphi(\omega)\Big|_{\omega=\omega_0} \qquad = \text{group delay} \qquad = \text{GD}(\omega_0)$$

$$D_2 = \partial_\omega^2 \varphi(\omega)\Big|_{\omega=\omega_0} \qquad = \text{group delay dispersion} = \text{GDD}(\omega_0)$$

$$D_3 = \partial_\omega^3 \varphi(\omega)\Big|_{\omega=\omega_0} \qquad = \text{third order dispersion} \quad = \text{TOD}(\omega_0)$$

$$D_4 = \partial_\omega^4 \varphi(\omega)\Big|_{\omega=\omega_0} \qquad = \text{forth order dispersion} \quad = \text{FOD}(\omega_0)$$

$$D_5, \ldots, D_\infty = \partial_\omega^{5\ldots\infty} \varphi(\omega)\Big|_{\omega=\omega_0} = \text{higher order dispersion}$$

Figure 2.4 shows the influence of the low order dispersion coefficients on the electric field of a laser pulse with a compressed pulse duration of 5 fs. It is apparent that dispersion can affect the CEP, the temporal delay, the instantaneous frequency ("chirp") and the pulse duration. Furthermore, the peak electric field strength is always reduced if any of the dispersion coefficients (with the exception of the group delay) is non-zero.

In many cases the described Taylor expansion is very useful as higher-order dispersion coefficients are often small and hence the spectral phase can be well approximated by a low-order polynomial allowing for an analytical treatment of dispersion effects. However, for broad spectral bandwidths and more complex phase curves $\varphi(\omega)$ the practicability of the Taylor expansion reaches its limit and a numerical treatment can be the better choice.

For visualization, instead of the phase $\varphi(\omega)$ often the frequency dependent group delay[3] $\text{GD}(\omega) = \partial_\omega \varphi(\omega)$ is plotted as it is the more intuitive quantity: sloppily speaking the GD gives the relative temporal delay of each frequency ω. Hence, a constant GD for all ω indicates a perfectly compressed pulse, a constant slope of the GD curve indicates a linear chirp and so on.

For most experiments it is crucial to work with compressed or specifically chirped pulses. Furthermore, it is common in laser amplifier chains to temporally stretch the optical pulses before amplification and to re-compress them afterwards. Therefore dispersion control plays a key role in ultrashort laser systems as will be briefly discussed in the following.

2.1.5 Chirped Pulse Amplification and Dispersion Control

The concept of chirped pulse amplification (CPA) has been first demonstrated by Strickland and Mourou in 1985 [17] and addresses the problem of laser induced damage of optical components in high energy amplifiers: by stretching the pulses to be amplified in time, the intensity can be reduced by orders of magnitude which allows a safe operation of the amplifier chain. In a subsequent compressor that is matched to

---

[3]Not to be confused with the dispersion coefficient for the group delay $D_1 = \text{GD}(\omega_0)$ which is the temporal delay of the carrier frequency ω_0 and hence a single value.

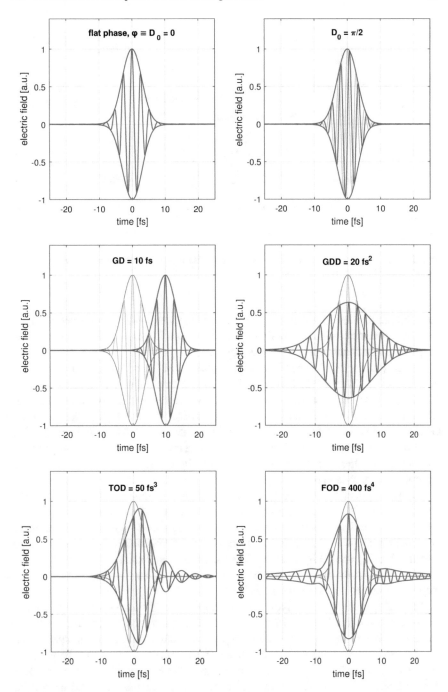

Fig. 2.4 Effect of different dispersion coefficients on the temporal waveform of a 5 fs pulse. The specified coefficients are evaluated at the carrier frequency $\omega_0 = 2.35$ PHz ($\lambda_0 = 800$ nm)

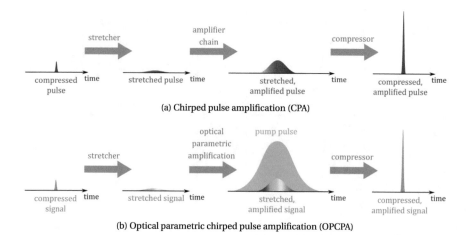

(a) Chirped pulse amplification (CPA)

(b) Optical parametric chirped pulse amplification (OPCPA)

Fig. 2.5 Illustration of the concepts of **a** chirped pulse amplification and **b** optical parametric chirped pulse amplification

the stretcher and to the additional dispersion of the beam path, the amplified pulses are re-compressed to finally obtain high-energy, high-intensity pulses. A schematic illustration of the concept is shown in Fig. 2.5a.

Optical parametric chirped pulse amplification (OPCPA, cf. Fig. 2.5b) transfers the idea of CPA to systems based on optical parametric amplification (OPA), a nonlinear amplification process that will be later explained in more detail. In contrast to classical CPA, the reason for stretching and re-compression of the "signal" pulses in the OPCPA scheme is usually not to prevent laser damage but to match their pulse duration with the duration of the typically orders of magnitude longer "pump" pulses. By this means, the extraction of pump energy by the signal is enhanced and the amplification gain is maximized [18].

For the application of these CPA concepts and to compensate dispersion introduced by media it is necessary to gain control over the individual dispersion coefficients of a pulse. To this end, a large variety of tools has been developed with the most prominent examples being prism and grating stretchers/compressors [19], chirped mirrors [20, 21], spatial light modulators (SLM) [22] and acousto-optic modulators (AOM) [23]. An excellent introduction to ultrashort pulse shaping using some of these instruments can be found in [24].

The usability of the respective tools depends very much on the specific application and often a combination of different schemes is implemented in one laser system. Table 2.3 gives a quick overview about typical advantages and disadvantages of the schemes without claiming validity for all possible cases (depending on the bandwidth, the wavelength region, and other factors, the performance might be better or worse than indicated). In the PFS system, all of the listed schemes are used as will be described in later chapters.

Table 2.3 Capabilities of different dispersion control schemes

Dispersion control scheme	High transmission	Suitable for high intensities	Compens. of higher order dispersion	Tunability	Large phase control window	Typical application
Prism stretcher/compressor	+	-	0	0	0	broadband CPA
Grating stretcher/compressor	0	+	-	0	+	CPA with large, high-energy beams
Chirped mirrors	+	+	+	-	-	ultrabroadband pulse compression
Liquid crystal spatial light modulator	0	-	+	+	-	fine control of dispersion, installed before amplification
Acousto-optic modulator	-	-	+	+	-	fine control of dispersion, installed before amplification

2.2 Nonlinear Processes

Having discussed diffraction and dispersion in the previous sections, the last term in the FOPE (Eq. (2.17)) that needs to be treated is the nonlinear one on the right-hand-side. Besides the frequency-dependent prefactor, this term contains the already defined nonlinear polarization which is given by

$$P_i^{(\text{NL})} = \varepsilon_0 \left(\chi_{ijk}^{(2)} E_j E_k + \chi_{ijkl}^{(3)} E_j E_k E_l + \cdots \right) \tag{2.42}$$

where the indices i, j, k denote the polarization components of the electric field and take the values x, y or z. For a more comprehensive notation we use the Einstein sum convention, i.e.:

$$\chi_{ijk}^{(2)} E_j E_k \,\hat{=}\, \sum_{j,k=x,y,z} \chi_{ijk}^{(2)} E_j E_k = \chi_{ixx}^{(2)} E_x E_x + \chi_{ixy}^{(2)} E_x E_y + \cdots \tag{2.43}$$

It is obvious that the nonlinear polarization $P_i^{(\text{NL})}$ is created by the combined interaction of different electric field components. During propagation in accordance with the FOPE this leads to an exchange of energy between the fields or the generation of new fields. The coupling strength between the components in this interaction is given by the material-specific nonlinear susceptibility tensors $\chi^{(n)}$ that will be discussed in Sect. 2.3.4 in more detail.

Depending on the number of fields that interact with each other (or with themselves) the induced nonlinear processes are referred to as second-order processes, third-order processes and so on. In the following the most important of these processes will be described.

2.2.1 Second-Order Processes

To understand how second-order nonlinear processes lead to the generation of new frequencies let us consider an electric field

$$E_i(z, t) = A_i(z)\, e^{i(k_i z - \omega_i t)} + \text{c.c.} \tag{2.44}$$

where each polarization direction $i \in \{x, y, z\}$ contains a distinct field amplitude $A_i(z)$ and frequency ω_i. The second-order nonlinear polarization is then given by:

$$
\begin{aligned}
P_i^{(2)} &= \varepsilon_0\, \chi_{ijk}^{(2)}\, E_j(z, t)\, E_k(z, t) \\
&= \varepsilon_0\, \chi_{ijk}^{(2)} \left(A_j(z)\, e^{i(k_j z - \omega_j t)} + \text{c.c.} \right) \left(A_k(z)\, e^{i(k_k z - \omega_k t)} + \text{c.c.} \right) \\
&= \varepsilon_0\, \chi_{ijk}^{(2)} \left(\underbrace{A_j(z)\, A_k(z)\, e^{i(k_j + k_k)z - i(\omega_j + \omega_k)t} + \text{c.c.}}_{\text{SFG}} \right. \\
&\qquad\qquad \left. + \underbrace{A_j(z)\, A_k^*(z)\, e^{i(k_j - k_k)z - i(\omega_j - \omega_k)t} + \text{c.c.}}_{\text{DFG}} \right)
\end{aligned}
\tag{2.45}
$$

where $A_k^*(z)$ is the complex conjugate of $A_k(z)$. As one can see from the time-dependent exponential terms, the nonlinear polarization oscillates at two new central frequencies:

- *Sum-frequency generation (SFG)* produces:

$$\textbf{SFG:} \quad \omega_j + \omega_k \rightarrow \omega_{\text{SF}} \tag{2.46}$$

In the special case that both input frequencies are identical ($\omega_j = \omega_k = \omega$), the process is called *second-harmonic generation (SHG)*:

$$\textbf{SHG:} \quad \omega + \omega \rightarrow \omega_{\text{SH}} = 2\,\omega \tag{2.47}$$

- For *difference-frequency generation (DFG)* it is common to name the three involved frequencies *pump*, *signal* and *idler*. Accordingly, the newly generated idler frequency ω_i is produced by:

$$\textbf{DFG:} \quad \omega_p - \omega_s \rightarrow \omega_i \tag{2.48}$$

The DFG process is often also referred to as *optical parametric amplification (OPA)* with the signal beam being amplified by the pump. Hence, the interaction can alternatively be written as:

$$\textbf{OPA:} \quad \omega_p \rightarrow \omega_s + \omega_i \tag{2.49}$$

Physically both processes are identical. The chosen designation generally depends on which beam is further used for experiments: the newly generated idler ("DFG") or the amplified signal beam ("OPA").

SFG and DFG are widespread techniques to convert light from one spectral region to another and allow to access different wavelength bands with just one primary light source. OPA on the other hand is used to transfer energy from the pump to the signal beam and to generate by this means high-energy pulses in spectral regions or with broad bandwidths that could not be accessed by classical laser amplification. As OPA is one of the most important processes for the experiments in this work, we will have a more detailed look onto it in the following.

Simple analytical simulation of optical parametric amplification

In order to derive a simple analytical model for OPA, we take only the DFG-term from Eq. (2.45) and omit the complex conjugate:

$$P_{\text{OPA},i}^{(2)+} = \varepsilon_0 \, \chi_{ijk}^{(2)} \, A_j \, A_k^* \, e^{i(k_j - k_k)z - i(\omega_j - \omega_k)t} \tag{2.50}$$

Restricting ourselves to the OPA interaction, we can furthermore substitute[4] the tensor $\chi_{ijk}^{(2)}$ by the scalar value d_{eff} from literature [1, 25] and replace the indices i, j, k in accordance with the pump-signal-idler nomenclature:

$$P_{\text{OPA},i}^{(2)+} = 4 \, \varepsilon_0 \, d_{\text{eff}} \, A_p \, A_s^* \, e^{i(k_p - k_s)z - i\omega_i t} \tag{2.51}$$

Using the ansatz

$$E_i^+(z, t) = A_i(z) \, e^{ik_i z - i\omega_i t} \tag{2.52}$$

for the electric field of the idler we can insert Eqs. (2.51) and (2.52) into the FOPE Eq. (2.17). For infinite plane waves the diffraction term vanishes and as we defined our electric fields to be monochromatic there is no dispersion. Hence we can write in the time domain:

$$\left[\partial_z - i k_i \right] E_i^+(z, t) = \frac{i \omega_i^2}{2 \varepsilon_0 c^2 k_i} P_{\text{OPA},i}^{(2)+}(z, t)$$

$$\Leftrightarrow \quad \left(\partial_z A_i(z) \right) e^{ik_i z - i\omega_i t} = \frac{2 \, i \, \omega_i^2 \, d_{\text{eff}}}{c^2 \, k_i} A_p(z) A_s^*(z) \, e^{i(k_p - k_s)z - i\omega_i t}$$

$$\Leftrightarrow \quad \partial_z A_i(z) = \frac{2 \, i \, \omega_i^2 \, d_{\text{eff}}}{c^2 \, k_i} A_p(z) A_s^*(z) \, e^{i\Delta k z} \tag{2.53}$$

where

$$\Delta k = k_p - k_s - k_i \tag{2.54}$$

is the **wavevector mismatch** and $\Delta k \, z$ is the **phase mismatch**. Equation (2.53) describes the evolution of the idler amplitude during propagation along z. The cor-

---

[4]This substitution adds a factor of two due to the historical definition of d_{eff} and another factor of two due to the permutation symmetry of $\chi_{ijk}^{(2)}$. The relation between d_{eff} and $\chi_{ijk}^{(2)}$ will be further explained in Sect. 2.3.4.

responding differential equations for signal and pump are respectively

$$\partial_z A_s(z) = \frac{2 \, i \, \omega_s^2 \, d_{\text{eff}}}{c^2 \, k_s} \, A_p(z) A_i^*(z) \, e^{i\Delta kz} \tag{2.55}$$

$$\partial_z A_p(z) = \frac{2 \, i \, \omega_p^2 \, d_{\text{eff}}}{c^2 \, k_p} \, A_s(z) A_i(z) \, e^{-i\Delta kz} \tag{2.56}$$

yielding all together the well known set of coupled PDEs for OPA [1]. Assuming $A_p \approx$ const. (i.e. neglecting pump depletion), $A_s(z{=}0) \ll A_p(z{=}0)$ and $A_i(z{=}0) = 0$, this set of PDEs can be solved analytically [2, 26] resulting in an expression for the parametric gain G of the signal intensity I_s:

$$\frac{I_s(z = L)}{I_s(z = 0)} = 1 + G = 1 + (gL)^2 \, \frac{\sinh^2\left(\sqrt{(gL)^2 - (\Delta k \, L/2)^2}\right)}{(gL)^2 - (\Delta k \, L/2)^2} \tag{2.57}$$

where L is the length of the medium and g is the gain coefficient:

$$g = 4\pi d_{\text{eff}} \sqrt{\frac{I_p}{2\varepsilon_0 n_p n_s n_i c \lambda_s \lambda_i}} \tag{2.58}$$

with the pump intensity I_p, the refractive indices n_p, n_s and n_i, and the wavelengths λ_s and λ_i.

Depending on the values of g, L and Δk there are three limit cases:

$$\Delta k \gg g \Rightarrow G \approx (gL)^2 \, \text{sinc}^2(\Delta k L/2) \tag{2.59}$$

$$\Delta k \ll g \text{ and } gL \ll 1 \Rightarrow G \approx (gL)^2 \tag{2.60}$$

$$\Delta k \ll g \text{ and } gL \gg 1 \Rightarrow G \approx \frac{1}{4} e^{2gL} \tag{2.61}$$

Figure 2.6 illustrates these cases with a few exemplary curves for different gain coefficient and wavevector mismatch values: The oscillatory behavior ("**back-conversion**") from Eq. (2.59), the quadratic increase from Eq. (2.60) at small L values and the exponential gain from Eq. (2.61) for small Δk and large L are clearly visible. It becomes obvious from the plots that a small Δk is very beneficial to achieve a high gain. Note that while in theory the negative effect of any $\Delta k \neq 0$ can be compensated with a larger gain coefficient g, in practice this is not feasible: Typically the only way to increase g is by increasing the pump intensity I_p and therefore the damage threshold of the medium and the onset of other nonlinear effects set a limit to this measure. Hence, minimizing Δk in the first place is a key requirement for efficient OPA as will be described in Sect. 2.3.3.

We will see that for broadband signal pulses a value of $\Delta k = 0$ cannot simultaneously be achieved for all wavelengths and therefore a wavelength-dependent phase mismatch remains. Besides the discussed implications for the parametric gain, this

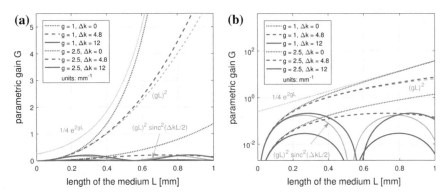

Fig. 2.6 a Simulated gain curves for monochromatic OPA as a function of distance L for different values of wavevector mismatch Δk and gain coefficient g. Grey curves show the limit cases for $g = 2.5$ and $\Delta k = 12$. **b** Same plot on logarithmic y-scale

mismatch also affects the spectral phase of the amplified signal pulses. The difference between amplified and unamplified signal phase is known as optical parametric phase or **OPA phase**. Under the assumption of negligible pump depletion it can be analytically expressed as [27]:

$$\varphi_{\text{OPA}} = -\frac{\Delta k L}{2} + \tan^{-1}\left(\frac{\Delta k L}{2} \frac{\tanh\left(\sqrt{(gL)^2 - (\Delta k L/2)^2} \right)}{\sqrt{(gL)^2 - (\Delta k L/2)^2}} \right) \tag{2.62}$$

In Fig. 2.7 this OPA phase is shown as a function of the parametric gain G for different phase mismatch values $\Delta k L$. As one can see from Fig. 2.7a, the magnitude of the OPA phase increases with higher gain and phase mismatch (but with a negative sign). To illustrate the functional interrelation, in Fig. 2.7b we plotted the OPA phase in units of the phase mismatch. It is apparent that for high gain values there is an almost constant ratio between the two that slowly converges towards -0.5 for $G \to \infty$. In practice this means that after high gain amplification, the signal pulses feature an additional phase $\varphi_{\text{OPA}}(\omega)$ which is negative proportional to the phase mismatch $\Delta k(\omega) L$ of the OPA process. Taking into account and compensating this additional phase is vital for a full temporal compression of the amplified pulses (an example can be found in [28]).

It should be pointed out that the analytical description of OPA here does only treat the interaction of three monochromatic waves in the case of low efficiency (non-depleted pump). So while the considerations provide some useful insight into the mechanism and scaling, a thorough treatment of broadband and highly efficient OPA (together with other nonlinear interactions) requires more sophisticated and numerical tools as will be described in Sect. 2.4.

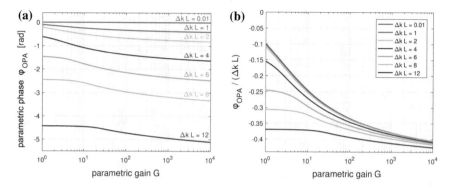

Fig. 2.7 **a** Simulated OPA phase as a function of parametric gain G for different phase mismatch values $\Delta k L$. Plot **b** shows again the OPA phase but this time in units of $\Delta k L$ to illustrate the tendency towards a constant ratio between OPA phase and phase mismatch for large gain values

2.2.2 Third-Order Processes

Since there exists a large variety of third-order processes we will limit ourself here to the effects most relevant for this work. These effects are ***self-focusing***, ***self-phase modulation*** (***SPM***), ***cross-phase modulation*** (***XPM***) and ***cross-polarized wave generation*** (***XPW***) and all of them can be symbolically described by a four-wave-mixing process:

$$\textbf{self-focussing, SPM, XPM, XPW:} \quad \omega_1 + \omega_2 - \omega_3 \rightarrow \omega_4 \qquad (2.63)$$

In the following we will describe the origin and physical consequences of these mixing processes.

Self-focusing, self-phase modulation and self-steepening

Let us assume a pulse propagates in z-direction with a linearly polarized electric field along the x-axis:

$$E_x(\mathbf{r}, t) = \mathcal{E}_x(\mathbf{r}, t) \cos(\omega t)$$
$$= \frac{\mathcal{E}_x(\mathbf{r}, t)}{2} \left(e^{-i\omega t} + \text{c.c.} \right) \qquad (2.64)$$

where the vector $\mathbf{r} = (x, y, z)$ accounts for a spatially dependent field strength. The third-order nonlinear polarization along the same axis is then given by:

$$P_x^{(3)} = \varepsilon_0 \, \chi_{xxxx}^{(3)} E_x^3(\boldsymbol{r}, t)$$

$$= \frac{\varepsilon_0}{8} \chi_{xxxx}^{(3)} \left[\mathcal{E}_x^3(\boldsymbol{r}, t) \left(e^{-i(\omega+\omega+\omega)t} + \text{c.c.} \right) + 3\mathcal{E}_x^3(\boldsymbol{r}, t) \left(e^{-i(\omega+\omega-\omega)t} + \text{c.c.} \right) \right]$$

$$= \frac{\varepsilon_0}{8} \chi_{xxxx}^{(3)} \left[\mathcal{E}_x^3(\boldsymbol{r}, t) \left(e^{-i3\omega t} + \text{c.c.} \right) + 3\mathcal{E}_x^3(\boldsymbol{r}, t) \left(e^{-i\omega t} + \text{c.c.} \right) \right]$$

$$= \frac{\varepsilon_0}{4} \chi_{xxxx}^{(3)} \left[\underbrace{\mathcal{E}_x^3(\boldsymbol{r}, t) \cos(3\omega t)}_{\text{THG}} + \underbrace{3\mathcal{E}_x^2(\boldsymbol{r}, t) E_x(\boldsymbol{r}, t)}_{\text{self-focusing, SPM}} \right] \tag{2.65}$$

As one can see, the nonlinear polarization contains one term that oscillates at 3ω (**third-harmonic generation, THG**) and another that reproduces the original field $E_x(\boldsymbol{r}, t)$, however, modulated by its intensity $I_x(\boldsymbol{r}, t) \sim \mathcal{E}_x^2(\boldsymbol{r}, t)$. While THG—in the same way as SHG – transfers energy from the fundamental frequency range into another spectral band, the second term modifies the fundamental pulse itself. This can be seen, when inserting $P_x^{(3)}$ into the FOPE (Eq. (2.17)) where this second term adds an intensity-dependent phase to the original field $E_x(\boldsymbol{r}, t)$ with important consequences:

(a) For a beam with *spatially* varying intensity $I = I(x, y)$ (e.g. a Gaussian beam) the additional phase leads for the common case of $\chi^{(3)} > 0$ to a slower propagation of the pulse wavefront in regions with high intensity compared to regions with low intensity resulting in a bent wavefront. Depending on the shape of the intensity distribution this can cause a focusing of the entire beam or of small intense regions ("hot spots") and is referred to as **self-focusing** [29].

(b) If the intensity varies over *time* (which is the case by definition for all short laser pulses) the introduced nonlinear phase is also time-dependent which implies (recalling the definition $\omega_{\text{inst}} = -d\Phi/dt$) the generation of new frequencies. This modification is called **self-phase modulation (SPM)**.

Additionally, the high-intensity temporal peak propagates slower than the less intense wings which is the temporal equivalent to the spatial self-focusing and results in a steepening of the pulse envelope at the trailing edge of the pulse (**self-steepening**) [30]. The effect of SPM and self-steepening is shown in Fig. 2.8 where we simulated the third-order nonlinear propagation of a pulse in a dispersion-free medium. The apparent spectral broadening induced by the combination of both effects can be experimentally exploited for **supercontinuum generation** in gas-filled hollow core fibers or in bulk material as we will show in Sect. 3.1.

The $\chi^{(3)}$-nonlinearity leading to self-focusing, SPM and self-steepening is often referred to as intensity-dependent **nonlinear refractive index $n(\lambda, I)$**:

$$n(\lambda, I) = n_0(\lambda) + n_2(\lambda) \cdot I + \cdots \tag{2.66}$$

where higher orders are usually neglected. The nonlinear refractive index provides an intuitive explanation of the additional phase delay introduced by high intensities and can be derived from $\chi^{(3)}$ [31]:

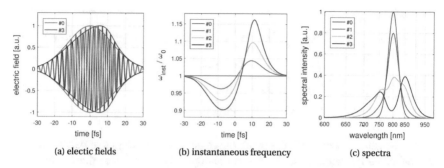

(a) electic fields   (b) instantaneous frequency   (c) spectra

Fig. 2.8 Simulation of SPM and self-steepening for 20 fs-pulses with different intensity (increasing from #0 to #3) propagating in an artificial medium featuring $\chi^{(3)}$-nonlinearity but no dispersion. **a** shows the electric fields of the two extreme cases where the modulated oscillation frequency and amplitude steepening towards the trailing edge of the high-intensity pulse are clearly visible. **b** shows the variation of the instantaneous frequency with time across the different pulses. In **c** the increasing spectral broadening can be seen

$$n_{2,i} = \frac{3}{4n_0^2\varepsilon_0 c}\chi_{iiii}^{(3)} \tag{2.67}$$

The nonlinear refractive index of gases is typically of the order of $10^{-19}\,\frac{m^2}{W}$ while for glasses and common nonlinear crystals it is of the order of $10^{-16}\,\frac{m^2}{W}$. Tabulated values for different materials can be found for example in [1, 2].

The nonlinear refractive index is often used to determine the so-called **B-integral**:

$$B = \frac{2\pi}{\lambda_0}\int_0^L n_2\,I(z)\,\mathrm{d}z \tag{2.68}$$

This integral measures the accumulated nonlinear phase shift after a propagation length L in units of one oscillation of the central wavelength λ_0. Hence, it provides an estimate for the impact of self-focusing and SPM on pulse shape and propagation. For beam transport these effects are usually unwanted and the B-integral is kept as small as possible e.g. by avoiding transmission of high-intensity pulses through glass substrates or by propagation in vacuum instead of air. In general, a B-integral value smaller than one is assumed to be tolerable.

Cross-phase modulation and cross-polarized wave generation
In contrast to self-focusing and SPM where the four interacting fields are polarized along the same direction and belong to a single beam, *cross-phase modulation* (*XPM*) denotes the case of third-order interactions between fields that belong to different beams or are polarized along different directions. The physical consequences are in fact quite similar to self-focusing and SPM with the difference that the nonlinear phase induced by one beam can be imprinted onto the other when the beams overlap in space and time [32]. Note that XPM represents a direct modulation of the

phase via $\chi^{(3)}$ and should not be confused with the situation when self-focusing and SPM in one of the beams affects the other via $\chi^{(2)}$ processes such as OPA [33].

Cross-polarized wave generation (XPW) is known as an $\chi^{(3)}$-effect where the electric field along one polarization direction not only modulates the phase of a cross-polarized field but even creates this field [34]. The exemplary generation of a y-polarization $P_{y,\text{XPW}}^{(3)}$ by an x-polarized field $E_x(r, t)$ can be written as:

$$P_{y,\text{XPW}}^{(3)} = \frac{3}{4} \varepsilon_0 \chi_{yxxx}^{(3)} \mathcal{E}_x^2(r, t) E_x(r, t) \tag{2.69}$$

The implementation of XPW generation into laser systems has become very popular in the past years [35–41] as it exhibits two attractive features:

(a) Since XPW generation is a third-order process, the envelope of the newly generated pulse is proportional to $P_{y,\text{XPW}}^{(3)}$ and therefore to the third power of the input pulse envelope:

$$\mathcal{E}_{y,\text{XPW}}(r, t) \sim \mathcal{E}_x^3(r, t) \tag{2.70}$$

Hence, in the ideal case of perfectly compressed input pulses, negligible dispersion, no saturation and absence of other nonlinear effects, XPW generation leads to a *pulse shortening* by $1/\sqrt{3} \approx 0.58$ which implies a *spectral broadening* of $\sqrt{3}$. Compared to spectral broadening by SPM this technique has the advantage that the broadening does not depend on input intensity. Thus, it can be used in cases where a high conversion efficiency (which *does* depend on intensity) is not required but pulses are too long for broadening by SPM or their phase would be critically distorted by the process.

(b) The second interesting feature of XPW generation is the fact that because of the high nonlinearity of the effect, any weak temporal artifacts (pre-/post-pulses) of the input field are strongly suppressed in the cross-polarized output. Hence, the contrast of the newly generated field is significantly enhanced: the process acts as a *temporal pulse cleaner*.

In Fig. 2.9 the simulation results for XPW generation by a 20 fs input pulse are shown. The spectral broadening as well as the temporal cleaning are clearly visible. Experimentally we will take advantage of both effects to produce high-contrast seed pulses for OPA as will be described in Sect. 4.2.

2.3 Nonlinear Media

Generally speaking, nonlinear processes in dielectric media become important if the electric field amplitude of an optical pulse is of the order of the atomic electric field strength E_{atom}. For the hydrogen atom for example this corresponds to a field of $E_{\text{atom}} \approx 5 \times 10^{11} \text{V/m}$. Hence, one can expect that the second-order susceptibility $\chi^{(2)}$ is of the order of $1/E_{\text{atom}} \approx 2 \text{ pm/V}$, the third-order susceptibility $\chi^{(3)}$ is of the

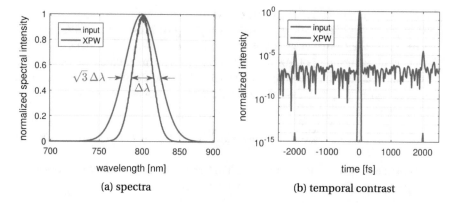

(a) spectra   (b) temporal contrast

Fig. 2.9 Simulation of spectral broadening and temporal cleaning for a pulse that propagates in an artificial non-dispersing medium with a $\chi^{(3)}$ nonlinearity supporting exclusively XPW generation. **a** Broadened spectrum of the cross-polarized wave compared to the input spectrum. **b** Contrast enhancement (logarithmic scale) by nonlinear suppression of temporal artifacts in the input field

Table 2.4 Tabulated values of the nonlinear susceptibility tensor components for selected crystals. For conversion between the contracted notations d_{il} and c_{im} and the genuine tensors $\chi^{(2)}_{ijk}$ and $\chi^{(3)}_{ijkl}$ see Table A.1 in the appendix

Material	Susceptibility tensor components		
BaF$_2$, $\chi^{(3)}$	$\chi^{(3)}_{xxxx} = 159\,\mathrm{pm}^2/\mathrm{V}^2$	$\sigma = \dfrac{\chi^{(3)}_{xxxx} - \left[\chi^{(3)}_{xxyy} + 2\chi^{(3)}_{xyyx}\right]}{\chi^{(3)}_{xxxx}} = -1.08$	[42]
BBO, $\chi^{(2)}$	$d_{23} = 2.2\,\mathrm{pm/V}$	$d_{31} = 0.16\,\mathrm{pm/V}$	[43]
BBO, $\chi^{(3)}$	$c_{11} = 500\,\mathrm{pm}^2/\mathrm{V}^2$	$c_{10} = -24\,\mathrm{pm}^2/\mathrm{V}^2$	
	$c_{16} = 147,\,\mathrm{pm}^2/\mathrm{V}^2$	$c_{33} = -535\,\mathrm{pm}^2/\mathrm{V}^2$	[44]
LBO, $\chi^{(2)}$	$d_{31} = d_{15} = 0.85\,\mathrm{pm/V}$	$d_{32} = d_{34} = -0.67\,\mathrm{pm/V}$	
	$d_{33} = 0.04\,\mathrm{pm/V}$		[45]

order of $4\,\mathrm{pm}^2/\mathrm{V}^2$ and so on. Indeed this estimate roughly matches the experimentally determined values of many dielectrics as can be seen from the susceptibility tensor components of a few exemplary materials listed in Table 2.4.

2.3.1 Isotropic Media

Whether a medium exhibits a certain type of nonlinearity depends very much on its atomic structure. Isotropic media like gases or glasses possess spherically symmetric binding potentials and therefore do not support even-order nonlinear processes such as SHG (except at interfaces). Hence, these media are typically used to trigger third-order effects like SPM. As air and glass substrates are often unavoidable components in laser systems also their unintended nonlinear impact (e.g. self-focusing) on high-

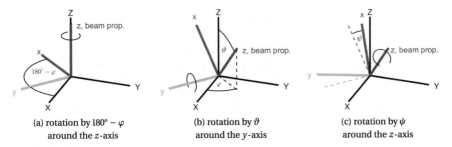

(a) rotation by 180° − φ
around the z-axis

(b) rotation by ϑ
around the y-axis

(c) rotation by ψ
around the z-axis

Fig. 2.10 Orientation of a beam inside a crystal: Successive rotation of the beam coordinate system (x, y, z) with respect to the principle axes of the crystal (X, Y, Z) by φ, ϑ and ψ. The z-axis indicates the direction of beam propagation, x and y are the directions of polarization

intensity pulses has to be considered. In the last section we have already introduced the B-integral (Eq. (2.68)) to quantify this impact.

2.3.2 Crystals

Crystals—in contrast to isotropic media—exhibit distinct symmetries in the atomic structure and especially may be non-centrosymmetric thus enabling even-order non-linear effects. An anisotropy of the crystalline structure generally implies also an anisotropy of the refractive index, i.e. the indices along the principle axes[5] of the crystal are different from each other. As a consequence, a laser beam that propagates through such a crystal "sees" different refractive indices depending on the orientation of the beam polarization with respect to the principle crystal axes. Hence, two orthogonally polarized electric fields in general are refracted differently at the crystal surface and propagate at different speed inside the material (**birefringence**).

If the electric field of a laser beam is not polarized along one of the principle axes of the crystal, the so-called "effective" refractive index has to be calculated. For this purpose one defines a beam coordinate system where the z-axis denotes the direction of propagation and x- and y-axis are the polarization directions[6] of the electric field. By convention, the orientation of this coordinate system with respect to the principle axes of the crystal is specified by the rotation angles φ, ϑ and ψ as shown in Fig. 2.10. The further procedure to determine the effective refractive indices n_x and n_y depends on the type of crystal.

Uniaxial crystals
Uniaxial crystals possess one optical axis Z with a refractive index different to the other two axes X and Y. For *negative* uniaxial crystals like **BBO** (beta-barium

---

[5]For orthorhombic crystals the principle axes correspond to the crystallographic axes. For non-orthogonal crystal structures such as hexagonal or triclinic crystals, however, principle and crystallographic axes do not coincide.

[6]For simplicity we consider in the following only linearly polarized beams.

borate, β-BaB$_2$O$_4$) or **DKDP** (potassium dideuterium phosphate, KD$_2$PO$_4$) the relation between the indices is $n_Z < n_X = n_Y$, whereas for *positive* uniaxial crystals it is $n_Z > n_X = n_Y$ [46]. The plane that contains the direction of beam propagation z and the optical axis Z of the crystal is termed "principle plane". A beam polarized perpendicular to this plane is called ***ordinary*** while a beam polarized in this plane is called ***extra-ordinary***. In Fig. 2.10b these polarization directions are represented by the y- and x-axis respectively. Hence, the ordinary refractive index is given by

$$n_{\mathrm{o}} = n_X \ (= n_Y),\tag{2.71}$$

whereas the extra-ordinary refractive index

$$n_{\mathrm{e}}(\vartheta) = \sqrt{\frac{\cos^2 \vartheta}{n_X^2} + \frac{\sin^2 \vartheta}{n_Z^2}}^{\,-1}\tag{2.72}$$

takes a value between n_X and n_Z that depends on the angle ϑ. Since $n_X = n_Y$, in uniaxial crystals a rotation around the optical axis by φ has no influence on both refractive indices.

For typical values of the "roll"-angle ψ (compare Fig. 2.10c) the beam polarization along the x-axis is either extra-ordinary or ordinary:

$$n_x = \begin{cases} n_e(\vartheta) & \text{for } \psi \in \{0°, 180°\} \\ n_o & \text{for } \psi \in \{90°, 270°\} \end{cases}\tag{2.73}$$

For the y-polarization with the effective refractive index n_y the assignment is vice versa. In the general case that ψ has any other than the specified values, x- and y-axis do not coincide with ordinary or extra-ordinary axes. As a consequence, the field components of a laser pulse E_x and E_y are decomposed into ordinary and extra-ordinary fields

$$E_o = E_x \sin \psi + E_y \cos \psi \tag{2.74}$$
$$E_e = E_x \cos \psi + E_y \sin \psi \tag{2.75}$$

with the respective refractive indices n_o and $n_e(\vartheta)$ as already specified. This decomposition is exploited for example in half-wave plates [47] where the input polarization splits into ordinary and extra-ordinary waves that propagate at different phase velocities. Hence, during propagation a phase delay between the two waves is created. When the two waves "recombine" at the exit surface of the wave plate this phase delay effectively results in a rotated polarization of the output beam with respect to the input.

Furthermore, the polarization splitting into ordinary and extra-ordinary components leads to the well known phenomenon of ***double refraction*** where objects appear as double images when observed through a birefringent crystal. Two physical effects are responsible for this phenomenon: first, according to Snell's law the polarization dependent refractive index leads to different angles of refraction at the crystal surface. And second, for the extra-ordinary beam (in contrast to the ordinary beam) the direction of energy flow does not coincide with the direction of the wavevector [43]. The angle ρ between these two directions is called **walk-off angle** and can be calculated as:

$$\rho(\vartheta) = \arctan\left(\left(\frac{n_o}{n_e}\right)^2 \tan\vartheta\right) - \vartheta \qquad (2.76)$$

where $n_e = n_e(\vartheta = 90°) = n_Z$. As a consequence of this walk-off, ordinary and extra-ordinary beams are spatially separated after propagation through a birefringent crystal. Remarkably, this effect also occurs for a perpendicular angle of incidence of the beam onto the crystal surface.

Biaxial crystals
In biaxial crystals the refractive indices n_X, n_Y, n_Z along the principle crystal axes are all different from each other. The assignment of the axes is by convention such that $n_Z > n_Y > n_X$. A biaxial crystal is called positive if n_Y is closer to n_X than to n_Z, otherwise it is called negative. The latter applies for example for **LBO** (lithium triborate, LiB_3O_5) [43].

Allowing for arbitrary values of φ and ϑ (i.e. an arbitrary beam orientation) all possible effective refractive indices in a biaxial crystal can be visualized by the so-called "refractive index ellipsoid" as shown in Fig. 2.11a. In contrast to uniaxial crystals it is for biaxial crystals in most cases not possible to identify an ordinary and an extra-ordinary polarization: in fact beams polarized along x or y are in general both extra-ordinary in the sense that their refractive indices depend on φ and ϑ and both beams exhibit walk-off.

However, in the special but commonly used case that the direction of propagation z lies in one of the principle planes XY, XZ or YZ, a biaxial crystal can be treated as quasi-uniaxial. This special orientation is obtained if φ (or ϑ) is $0°$ or a multiple of $90°$ and is shown for two exemplary cases in Fig. 2.11b, c. Consequently, n_y (or n_x) represents the ordinary refractive index and is identical to one of the principle indices while n_x (or n_y) represents the extra-ordinary index and takes a value between the other two principle indices as a function of ϑ (or φ). For the calculation of n_x and n_y the following formulas can be used:

$$n_x = \sqrt{\frac{\cos^2(\varphi)\cos^2(\vartheta)}{n_X^2} + \frac{\sin^2(\varphi)\cos^2(\vartheta)}{n_Y^2} + \frac{\sin^2(\vartheta)}{n_Z^2}}^{-1} \qquad (2.77)$$

$$n_y = \sqrt{\frac{\sin^2(\varphi)}{n_X^2} + \frac{\cos^2(\varphi)}{n_Y^2}}^{-1} \qquad (2.78)$$

Having described now in this section how the effective refractive indices in birefringent crystals depend on the relative orientation of principle crystal axes and beam

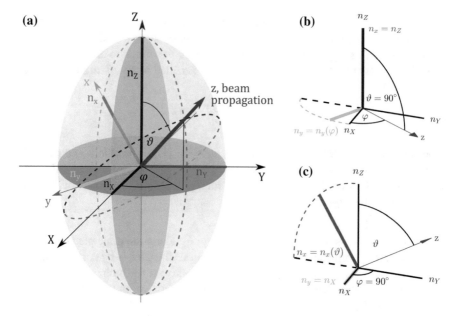

Fig. 2.11 **a** Refractive index ellipsoid of a biaxial crystal. n_X, n_Y, n_Z are the refractive indices along the principle crystal axes X, Y, Z and n_x, n_y are the effective refractive indices along the axes x, y of a beam coordinate system rotated by φ and ϑ relative to the crystal system. **b** and **c** show two examples of quasi-uniaxial orientations of a biaxial crystal

polarizations, we will discuss in the following how this property is used to achieve phase matching in nonlinear processes.

2.3.3 Critical Phase Matching

In general, the electric fields that take part in a nonlinear interaction cover different and/or broad wavelength bands. As we have already discussed, these different spectral components propagate at different phase velocities because of material dispersion. Hence, without further action the individual fields will be out of phase after a propagation distance equal to the coherence length $L_c = \pi / \Delta k$, where Δk is the wavevector mismatch of the interacting fields. This loss in coherence eventually leads to the reversal of a nonlinear interaction ("back-conversion") and thus reduces its efficiency. Using OPA as an example we have seen this effect already in Fig. 2.6.

In this respect, the birefringence of many nonlinear crystals is a very beneficial property as it allows to tune the effective refractive indices n_x and n_y by adjusting the orientation of the crystal. Since the wavevectors of the interacting fields depend on the index of refraction ($k = k_{vac} \cdot n$) by this means the wavevector mismatch Δk can be minimized. This method is called **critical phase matching** as opposed to other

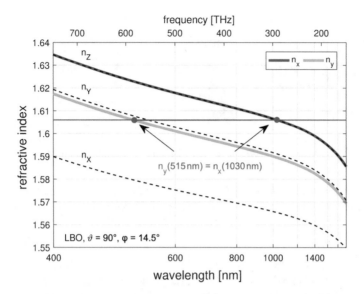

Fig. 2.12 Example for critical phase matching in an LBO crystal: by rotation of the crystal, the effective refractive indices n_X (1030 nm) and n_y (515 nm) are matched to support efficient SHG: 1030 nm $\rightarrow$ 515 nm. The principle refractive indices of LBO n_X, n_Y, n_Z are shown as dashed lines

phase matching techniques such as temperature phase matching [2] or quasi-phase-matching [48].

Figure 2.12 shows how the SHG-process 1030 nm $\rightarrow$ 515 nm in an LBO crystal can be phase matched by rotating the crystal such that the refractive index at 1030 nm on the x-axis (ordinary) and the index at 515 nm on the y-axis (extra-ordinary) are equal. For the calculation of n_x and n_y the Eqs. (2.77) and (2.78) can be used.

While this example shows that by crystal angle tuning it is possible to phase-match two discrete wavelengths, the frequency conversion and amplification of ultrashort laser pulses requires *simultaneous* phase matching for spectral bandwidths of up to several hundred nanometers. In the case of OPA with **collinear** pump, signal and idler beams this means that the wavevector mismatch

$$\Delta k = k_p - k_s - k_i = \frac{1}{c}\left(w_p\, n_p - w_s\, n_s - w_i\, n_i\right) \tag{2.79}$$

has to be zero (or close to zero) for all interacting frequencies yielding the condition

$$n_i\left(\omega_p - \omega_s\right) \overset{!}{=} n_p\,\omega_p - n_s\,\omega_s \qquad \forall\, \omega_p, \omega_s \tag{2.80}$$

where we used the conservation of energy $\omega_i = \omega_p - \omega_s$.

For most materials and spectral regions this condition cannot be fulfilled due to the unfavorable development of the refractive indices with ω. A common method to work around this issue is to introduce a **non-collinear angle** α between pump

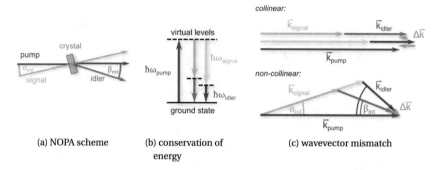

Fig. 2.13 Schematics of **a** the non-collinear interaction of pump, signal and idler, **b** conservation of energy of the involved photons for different signal/idler frequencies, **c** wavevector mismatch in collinear and non-collinear OPA for different signal/idler wavevectors

and signal beam (see Fig. 2.13a) as an additional degree of freedom [49]. To avoid confusion, in the following we will call the angle between pump and signal *inside* the interaction medium α_{int} whereas the angle *outside* the medium will be called α_{ext}. To describe the wavevector mismatch in such a ***non-collinear OPA (NOPA)*** scheme, Eq. (2.79) has to be modified to a vectorial equation:

$$\Delta k = k_{\mathrm{p}} - k_{\mathrm{s}} - k_{\mathrm{i}} \tag{2.81}$$

$$\Rightarrow \quad \Delta k = \left\|\Delta k\right\| = \sqrt{k_{\mathrm{p}}^2 + k_{\mathrm{s}}^2 - 2\,k_{\mathrm{p}}\,k_{\mathrm{s}}\cos\left(\alpha_{\mathrm{int}}\right)} - k_{\mathrm{i}} \tag{2.82}$$

Compared to collinear OPA, the free parameter α_{int} allows to reduce the mismatch Δk for different signal/idler wavevector pairs geometrically as depicted in Fig. 2.13c.

Figure 2.14 demonstrates numerically how such a NOPA scheme can be used to phase-match a broadband signal ($\lambda_{\mathrm{s}} = 680 - 1400\,\mathrm{nm}$) and a narrowband pump ($\lambda_{\mathrm{p}} = 515\,\mathrm{nm}$) in an LBO crystal. The chosen wavelengths correspond to the OPCPA design of the PFS system that will be described in the next chapter. Using the absolute value of the wavevector mismatch integrated over the full wavelength range as a cost function, Fig. 2.14a shows the dependence of the optimal phase-matching angle φ on the non-collinear angle α_{int} (white line). The wavelength-resolved wavevector mismatch for selected points on this curve is displayed in Fig. 2.14b where $\alpha_{\mathrm{int}} = 1.1°$ and $\varphi = 14.4°$ can be identified as the optimal parameter set to accomplish at the same time a small total mismatch and a broad phase-matching bandwidth. The dashed curves show the respective Δk values for this parameter set if the pump wavelength would be changed by $\pm 0.5\,\mathrm{nm}$ and hence emulate the slightly relaxed phase matching with a non-monochromatic pump. Figure 2.14c, d illustrate how the wavevector mismatch changes if one of the two angles is scanned while the other is fixed.

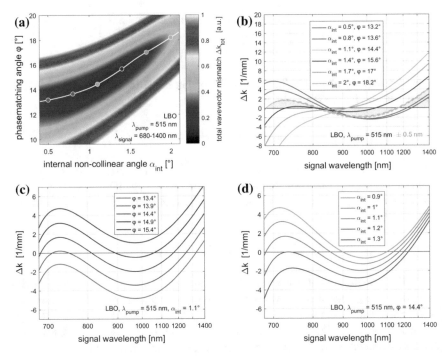

Fig. 2.14 Wavevector mismatch for broadband NOPA in an LBO crystal. (a) total mismatch $\Delta k_{tot} = \int |\Delta k(\omega_s)| \, d\omega_s$ for different combinations of α_{int} and φ. The white line indicates for each α_{int} the optimal angle φ to achieve minimal total mismatch. (b) Wavelength-resolved wavevector mismatch for selected combinations of α_{int} and φ. Dashed lines show mismatch for a ± 0.5 nm shifted pump wavelength emulating a broadband pump. (c) Scan of phase matching angle φ. (d) Scan of non-collinear angle α_{int}

2.3.4 Nonlinear Susceptibility Tensors

The nonlinear susceptibility tensors $\chi^{(2)}$, $\chi^{(3)}$, ..., $\chi^{(n)}$ that define the coupling strength between electric fields in a nonlinear interaction possess without further assumptions 3^n independent elements. However, by applying crystal and Kleinman symmetries this number can often be greatly reduced [1]. Additionally, it is common to use the contracted notation with the matrices d_{il} and c_{im} instead of $\chi_{ijk}^{(2)}$ and $\chi_{ijkl}^{(3)}$. Hence, often only few independent and non-zero elements remain (compare Table 2.4).

Similar to the effective refractive index, there is an effective susceptibility tensor that determines the nonlinear interaction for a beam propagating in a crystal under some angle. By using the rotation matrix

$$T(\varphi, \vartheta) = \begin{pmatrix} -\cos\varphi\cos\vartheta & -\sin\varphi\cos\vartheta & \sin\vartheta \\ \sin\varphi & -\cos\varphi & 0 \\ \cos\varphi\sin\vartheta & \sin\varphi\sin\vartheta & \cos\vartheta \end{pmatrix} \tag{2.83}$$

where the angles φ and ϑ are the same as defined in Fig. 2.10, one can calculate[7] the effective second-order nonlinear tensor $\chi^{(2)'}_{lmn}$:

$$\chi^{(2)'}_{lmn} = T_{li}\, T^{-1}_{jm}\, T^{-1}_{kn}\, \chi^{(2)}_{ijk} \tag{2.84}$$

where T^{-1} is the inverse of T. The dash in $\chi^{(2)'}_{lmn}$ indicates that the tensor is now given in the basis of the beam coordinate system (x, y, z) while the original tensor without dash is given in crystal coordinates (X, Y, Z). Accordingly, for the third order susceptibility holds:

$$\chi^{(3)'}_{mnpq} = T_{mi}\, T^{-1}_{jn}\, T^{-1}_{kp}\, T^{-1}_{lq}\, \chi^{(3)}_{ijkl} \tag{2.85}$$

To illustrate the idea of tensor transformation we will use the $\chi^{(2)}$-tensor of BBO (3m point group crystal structure) that has eleven non-zero components with three independent values, namely d_{22}, d_{31} and d_{33} [2]. As the value of d_{33} is very small it is usually neglected. Performing now the matrix multiplications of Eq. (2.84) with this tensor and taking a look at the component $\chi^{(2)'}_{xyy}$, one obtains:

$$\chi^{(2)'}_{xyy} = 2\left(d_{31}\,\sin\vartheta - d_{22}\,\sin 3\varphi\,\cos\vartheta\right) = 2\,d_{\text{eff}} \tag{2.86}$$

with the well-known d_{eff} value from literature [1]. In the typical case of $\varphi = 90°$ we can identify x as the extra-ordinary and y as the ordinary axis, thus $\chi^{(2)'}_{xyy}$ (like d_{eff}) is the coupling constant for the Type I interaction $o + o \rightarrow e$.

In summary, the matrix formalism allows to easily calculate the effective nonlinear susceptibility tensors for any crystal orientation. Using this formalism for the simulation of nonlinear interactions has the advantage over simulations based on a single d_{eff} value that it automatically considers all nonlinear coupling coefficients and therefore all possible interactions are included.

2.4 Numerical Simulations

In order to understand and optimize nonlinear processes, considerable efforts have been made to simulate the propagation and interaction of short light pulses. In particular OPA with its ability to generate ultra-broadband pulses with high energy has raised a great interest [1, 2, 18, 26, 50–52]. While analytical considerations such as presented in Sect. 2.2.1 help to identify the important quantities and constraints,

---

[7]For a derivation of Eq. (2.84) see Sect. A.2.2 in the appendix.

they generally have to make certain assumptions to be analytically solvable. These assumptions—e.g. regarding the spectral and temporal shape of the pulses, regarding dispersion, pump depletion, etc.—often do not reflect the conditions in real world systems.

Hence, numerical tools have been developed to overcome these limitations. The most prominent one is probably the SNLO software by Arlee Smith [53] that serves as an excellent database for nonlinear crystals and their properties and as a simulation tool for second-order interactions. However, as it lacks the possibility to include experimentally determined pulse features such as spectrum, spectral phase or spatial shape and as it does not allow to run batch parameter scans, it cannot be used for all applications. Furthermore, it is based on a single nonlinear coupling coefficient d_{eff} and hence does not include multiple second-order or any third-order interactions.

The SISYFOS package by Gunnar Arisholm [50] on the other hand is able to integrate experimental data or arbitrary pulse parameters and has been extensively used to simulate OPCPA systems such as PFS [52, 54, 55]. It constitutes a very powerful tool that contains most linear and nonlinear effects (except for XPM, XPW) and even allows the simulation of entire optical systems consisting of multiple components. A drawback of these capacities is that the tool is quite complex and being closed-source the code cannot be freely analyzed or extended.

Other tools such as the SOPAS code written by Christoph Skrobol in our group [55] contain some of the features of SISYFOS but also suffer from restrictions (such as allowing only for monochromatic pump pulses) that limit the applicability for the interaction of multiple broadband pulses.

As an extension to the available tools, we therefore developed a code specialized on the interaction of collinearly propagating ultra-broadband pulses including dispersion, linear absorption and all second- and third-order nonlinear effects. Written entirely in MATLAB, the code is easy to read, maintain and extend and allows the input and output of data at any time.

Since the goal is to simulate the interaction of broadband pulses that might spectrally overlap ("degenerate" interactions), the frequently used approach to distinguish beams by their spectral bands is not practicable. Furthermore, the limitation to certain wavelength bands would intrinsically suppress unpredicted mixing products generated e.g. by cascaded processes. Therefore we follow the more general approach (described for example in [56]) to distinguish pulses (or rather their electric fields) not by spectral regions but only by their polarization direction. As a starting point serves the FOPE where we limit ourself to infinite plane waves and therefore neglect the diffraction term:

$$\partial_z \tilde{E}_i(z, \omega) = ik_i \tilde{E}_i(z, \omega) + \frac{i\omega}{2\varepsilon_0 n_i c} \tilde{P}_{NL,i}(z, \omega)$$

$$\approx ik_i \tilde{E}_i(z, \omega) + \frac{i\omega}{2n_i c} \left(\chi_{ijk}^{(2)} \mathfrak{F}\{E_j(z, t) E_k(z, t)\} + \chi_{ijkl}^{(3)} \mathfrak{F}\{E_j(z, t) E_k(z, t) E_l(z, t)\} \right)$$

$$(2.87)$$

Higher order nonlinear susceptibility tensors like $\chi_{ijklm}^{(4)}$ are not considered (but could be included if necessary). Assuming that pulses propagate along the z-axis all indices in this equation can take the value x or y. Hence, Eq. (2.87) in fact defines two coupled PDEs describing the propagation and interaction of the electric fields E_x and E_y. This set of PDEs can be solved numerically in different ways. We chose to use the exponential Euler method [57] which solves a differential equation of the form

$$\partial_z y(z) = Ay + \mathcal{F}(y) \tag{2.88}$$

by

$$y_{s+1} = e^{A\,\Delta z}\,y_s - \frac{1 - e^{A\,\Delta z}}{A}\mathcal{F}(y_s) \tag{2.89}$$

where Δz is the appropriately chosen step size and s iterates until $s \cdot \Delta z$ reaches the predefined end position. Applying this method on Eq. (2.87) and calculating the effective refractive indices and susceptibility tensors as described in Sect. 2.3 (indicated by the dash symbols), one obtains:

$$\tilde{E}'_{m,s+1}(\omega) = e^{i\kappa_m \Delta z}\,\tilde{E}'_{m,s}(\omega) - \frac{1 - e^{i\kappa_m \Delta z}}{\kappa_m}\,\frac{\omega}{2n'_m c}\left(\chi_{mnp}^{(2)\prime}\,\mathfrak{F}\left\{E'_{n,s}(t)\,E'_{p,s}(t)\right\} + \right.$$
$$\left. \chi_{mnpq}^{(3)\prime}\,\mathfrak{F}\left\{E'_{n,s}(t)\,E'_{p,s}(t)\,E'_{q,s}(t)\right\}\right) \tag{2.90}$$

This set of equations (once more m, n, p, q can take the value x and y) corresponds to a 1D+time split-step-method, i.e. for each step s it switches between the frequency domain in which the linear propagation (dispersion and absorption) is calculated and the time domain in which the nonlinear interaction is performed. For a better numerical stability we use reduced wavenumbers $\kappa_m = k_m - k_0$, where k_0 is the wavenumber of the central frequency ("moving frame").

For a test of the fidelity of the exponential Euler code we compared it with SNLO ("PW-mix SP" method) by calculating with both tools the SHG efficiency of 30 fs pulses at $\lambda_0 = 800$ nm central wavelength in a BBO crystal. In our code we set $\chi^{(3)}$ to zero thus only $\chi^{(2)}$ effects were included. As can be seen in Fig. 2.15 the simulation results are quasi-identical demonstrating that the exponential Euler code in principle works (although experimentally the high efficiencies suggested by both simulations will hardly be achieved due to the onset of other nonlinear effects and optical damage of the BBO crystal).

As the stability of the code strongly depends on the step size Δz and the number of equidistant frequency points N_ω, these parameters have to be chosen carefully. In practice, it proved to be a good indication for sufficiently fine sampling if the sum of input and the sum of output energies of the simulated beams were identical (when absorption is switched off). Since the code does not include any lateral dimension, it is reasonably fast: at $N_{\Delta z} = 1000$ slices and $N_\omega = 2^{14}$ frequency points the computation takes less than two seconds on one core of an Intel® Xeon E3-1240.

2.5 Temporal Characterization of Ultrashort Light Pulses

To resolve a short phenomenon such as a light pulse in time, one always needs a comparably short reference that allows to temporally sample the phenomenon. In the case of ultrashort light pulses the shortest references at hand are in fact other light pulses or the original pulse itself. For this all-optical characterization of light pulses several methods have been developed that can mostly be assigned to three categories:

With the second-order *auto-correlation* technique a light pulse is measured by splitting it in a Michelson-interferometer setup into two duplicates which are then recombined and overlapped in time and space at a second-order nonlinear crystal [3]. Recording the intensity of the generated second-harmonic signal while temporally delaying one of the duplicates with respect to the other yields an auto-correlation signal over time. This signal can be used to determine the pulse duration of the original light pulse. If the pulse to be examined is combined with a known reference pulse (instead of creating and combining two duplicates), the technique is referred to as *cross-correlation*. The problem of auto- as well as cross-correlation is that while they provide a proper estimate for the pulse *duration*, the actual temporal *shape* cannot be unambiguously retrieved.

Interferometric methods represent the second category of pulse measurement techniques and take advantage of the fact that according to Eq. (2.33) a light pulse is fully characterized in time if its spectrum $S(\omega)$ and spectral phase $\varphi(\omega)$ are known.

Fig. 2.15 Comparison of the simulation results obtained by SNLO and the exponential Euler code. For the simulation 30 fs pulses with central wavelength $\lambda_0 = 800$ nm were used

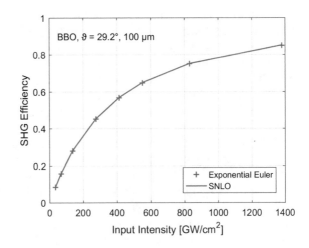

Using *Spectral Interferometry* (*SI*) [58], two pulses with a fixed temporal delay Δt between them are combined at a beam sampler and sent to a spectrometer. The measurement of the individual pulse spectra $S_1(\omega)$ and $S_2(\omega)$ is performed by blocking the other beam. To determine the spectral phase, one can use the modulations in the combined spectrum that result from the interference of the two pulses. From the mathematical expression for these modulations, given by

$$S_{\mathrm{mod}}(\omega) \propto \cos\Big(\varphi_1(\omega) - \varphi_2(\omega) - \omega\Delta t\Big), \tag{2.91}$$

one can infer that they provide only information about the *relative* spectral phase $\varphi_{\mathrm{rel}}(\omega) = \varphi_1(\omega) - \varphi_2(\omega)$ between the pulses. In particular for the case that a replica of the first pulse is used as the second one ($\varphi_1 = \varphi_2$), the relative phase term vanishes entirely. Therefore, SI in general does not allow the full characterization of a light pulse.

An established technique to work around this issue is *Spectral Phase Interferometry for Direct Electric-field Reconstruction* (*SPIDER*) [59]. The principle of SPIDER is to create two replicas of the pulse to be characterized, introduce a fixed temporal delay Δt between them, and mix them with a chirped copy of the original pulse in a nonlinear crystal. The sum-frequency pulses generated by this means exhibit a temporal offset Δt as well as a spectral offset $\Delta\omega$ with respect to each other. As a consequence, the interference spectrum of the sum-frequency pulses is given by

$$S_{\mathrm{mod}}(\omega) \propto \cos\Big(\varphi(\omega) - \varphi(\omega + \Delta\omega) - \omega\Delta t\Big). \tag{2.92}$$

Since Δt and $\Delta\omega$ can be determined from the experimental setup, this allows to calculate the absolute spectral phase $\varphi(\omega)$ of the input pulse (modulo a constant phase and group delay term). As SPIDER is furthermore intrinsically a single-shot method, it is widely used for pulse characterization and many variants have been developed [60].

A rather new but conceptually straight-forward modification of the SI principle to allow the measurement of the absolute spectral phase is the *Self-Referenced Spectral Interferometry* (*SRSI*) [61, 62]. It creates a reference pulse by XPW generation from the pulse to be characterized and measures the interference spectrum between the two. Due to the cubic nonlinearity of the XPW generation process, the reference pulse features a significantly smoothed spectral phase that can in most cases be considered to be flat. Hence, the measured relative phase between the interfering pulses represents approximately the absolute phase of the test pulse. Compared to SPIDER, the SRSI setup is simpler in design and the measured signal is located in the same frequency band as the input spectrum since no frequency conversion is involved. This makes SRSI for some applications the better choice.

The third category of pulse characterization methods is represented by variants of the *Frequency Resolved Optical Gating* (*FROG*) technique [3]. As the name suggests, the idea of FROG is to temporally crop the test pulse $E(t)$ with a gate $G(t)$ and to measure the spectrum of the resulting gated pulse as a function of the delay

Δt between test and gate pulse. This yields a signal of the form

$$S(\Delta t, \omega) \propto \left| \int E(t)\, G(t - \Delta t)\, e^{i\omega t}\, \mathrm{d}t \right|^2 , \qquad (2.93)$$

i.e. a 2D spectrogram, also known as the FROG trace. From this spectrogram the test and gate pulse can be iteratively retrieved with Principal Component Generalized Projections Algorithms (PCGPA) [63, 64].

Depending on what is used as a gate $G(t)$, one distinguishes different FROG types [65]. During the course of this work, three FROG devices were built and used for pulse characterization as will be shortly explained in the following.

2.5.1 Cross-Correlation FROG

In a cross-correlation FROG (XFROG) scheme, a typically compressed reference pulse is used as the gate pulse $G(t)$ which is non-collinearly mixed with the test pulse in a second-order nonlinear crystal. The resulting sum frequency is recorded as a function of the relative delay between the two pulses and yields the 2D spectrogram Fig. 2.16. shows a sketch of the home-built XFROG setup: gate and test pulse are sent to a flat bi-mirror, where one of the mirrors is mounted on a motorized delay stage. A subsequent spherical mirror focuses both beams into a thin BBO crystal to generate the sum frequency. The residual fundamental beams as well as the respective second-harmonics are blocked by a horizontal slit. A lens images the sum-frequency beam from the crystal onto the entrance of a spectrometer fiber for the acquisition of the spectrum.

The probably most beneficial property of the XFROG method is the usually straight-forward interpretability of the FROG trace: if the gate pulse is "well-behaved", i.e. if it does not exhibit strong amplitude modulations or phase distortions,

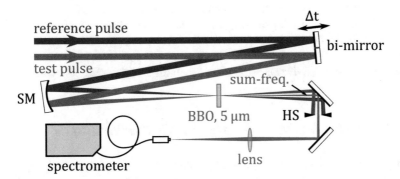

Fig. 2.16 Sketch of the home-built XFROG setup. SM = spherical mirror, HS = horizontal slit

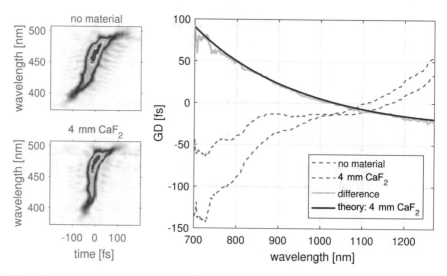

Fig. 2.17 Comparison of experimentally determined material dispersion from two XFROG measurements with theory. The FROG traces shown to the left are obtained by gating uncompressed broadband pulses (700–1300 nm) with compressed 25 fs reference pulses at 795 nm central wavelength. For the measurement of the lower trace, a 4 mm CaF_2 substrate was placed in the beam path of the broadband pulses. The main plot shows the respective retrieved GD curves and the difference between the two as well as the theoretical dispersion of CaF_2 calculated from the Sellmeier equation

the intensity distribution of the trace closely resembles the group delay curve $GD(\omega)$ of the test pulse. Hence, often no retrieval of the spectrogram is required to judge the chirp of the test pulse and to decide how dispersion has to be adjusted in order to achieve best compression.

As the temporal delay Δt is set by the mechanical movement of one of the bimirrors, it is possible to measure ultrashort (few femtoseconds) as well as comparably long or chirped pulses (tens of picoseconds) with our setup. This flexibility comes at a price which is the scanning nature of the device. Thus, shot-to-shot-stable test and gate pulses are required to obtain clean and meaningful FROG traces.

Exemplary traces are shown in Fig. 2.17 where we checked the performance of our XFROG setup by comparing measured material dispersion with theory. As one can see from the main plot, the difference in retrieved GD curves matches the theoretical expectation very well, demonstrating the capabilities of the XFROG method.

2.5.2 Single-Shot Second-Harmonic Generation FROG

A FROG variant that—different to the XFROG—does not rely on a reference pulse is the so-called second-harmonic generation FROG (SHG-FROG). As a gate it uses the test pulse itself and is thus a self-referencing method. Its spectrogram

$$S(\Delta t, \omega) \propto \left| \int E(t) \, E(t - \Delta t) \, e^{i\omega t} \, dt \right|^2 \qquad (2.94)$$

is always symmetric in time and has to be retrieved by PCGPAs to obtain information about the spectral phase of the test pulse (in contrast to the intuitive XFROG traces). A scanning SHG-FROG scheme would look identical to the XFROG version in Fig. 2.16 with the only difference that instead of the reference pulse a copy of the test pulse would be used at the input.

To make an SHG-FROG setup a single-shot device (**SS-SHG-FROG**), it is necessary to modify the setup such that the time delay is encoded along one spatial dimension and the spectrum along the other. In this way, a 2D sensor such as a CCD camera can capture the spectrogram in a single measurement [66]. Figure 2.18 shows the realization of such a scheme as a schematic layout.

In the top view in Fig. 2.18a the frequency encoding on the CCD can be seen. It is accomplished by focusing the input beam with a cylindrical mirror into a nonlinear crystal, re-collimating the generated second harmonic with two spherical mirrors and finally separating the wavelengths with a prism and a lens. A horizontal slit located in the image plane of the nonlinear crystal is used to adjust the spectral resolution and to remove out-of-focus artifacts.

In the side view in Fig. 2.18b, c the time encoding is shown. It is achieved by splitting the input beam into two halves with a bi-mirror and crossing these halves under an angle inside the nonlinear crystal (note that in this dimension the cylindrical mirror acts as a flat mirror). As can be seen from the magnified inset in Fig. 2.18c, this angle leads to a tilt between the pulse fronts of the two beam halves and thus to a varying delay across the beam profile. This spatio-temporal relation is imprinted onto the generated second-harmonic signal which is subsequently imaged by two spherical mirrors and a lens onto the CCD. A vertical slit removes the fundamental beams together with their respective second harmonics.

In order to be able to put correctly scaled axes to the acquired camera pictures, we calibrated our SS-SHG-FROG device spectrally with a set of narrowband dielectric filters and temporally with a double-pulse featuring a known delay. The spectral sensitivity was determined by using a compressed broadband pulse and comparing the time-integrated spectrum of the measured trace with the theoretical second-harmonic spectrum. Correcting for this sensitivity, finally the calibrated spectrogram is obtained. For all measurements described later in this work, we used the freely available FROG code from Rick Trebino's group [64] to retrieve the pulse properties from the spectrograms.

Figure 2.19 shows a test measurement with broadband pulses where we checked once more the consistency of measured and theoretical dispersion with a CaF_2 substrate. Some deviation of the GD curves is visible in the short wavelength region <900 nm where a reliable retrieval is difficult due to the low second-harmonic signal in this range in the measurement with the CaF_2 substrate. The overall agreement of theory and experiment, however, is quite good and allows the temporal characterization of ~ 5 fs pulses with our SS-SHG-FROG device.

(a) top view

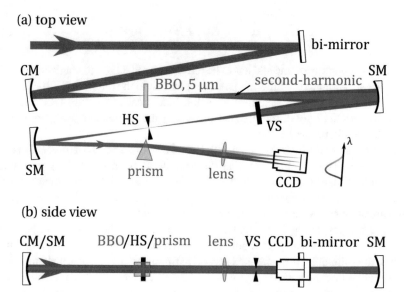

(b) side view

(c) pseudo side view

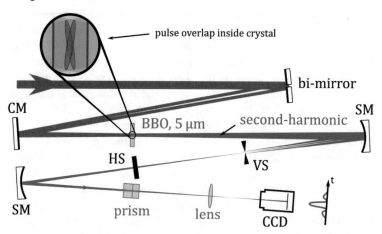

Fig. 2.18 Sketch of the home-built single-shot SHG-FROG setup: **a** shows the view from the top, **b** shows the view from the side and **c** shows a pseudo side view where the beam path is unfolded and drawn similar to (**a**) for a better visibility. CM = cylindrical mirror, SM = spherical mirror, VS = vertical slit, HS = horizontal slit

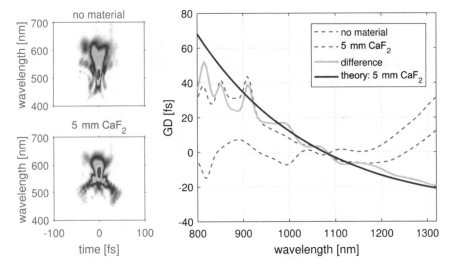

Fig. 2.19 Consistency check of experimentally determined material dispersion from two SS-SHG-FROG measurements and theory. To the left the two calibrated FROG traces are shown where an additional 5 mm CaF$_2$ substrate was placed in the beam path for the measurement of the lower trace. The main plot shows the respective retrieved group delay curves and the difference between the two as well as the theoretical dispersion of CaF$_2$ calculated from the Sellmeier equation

2.5.3 Transient-Grating FROG

The third FROG type which we set up to characterize our light pulses is a transient-grating FROG (TG-FROG) [67]. In contrast to the FROG types presented before, it exploits the *third*-order nonlinearity of a medium for the gating process. By crossing two intense light pulses under an angle in a $\chi^{(3)}$ medium, an interference pattern is created which locally alters the refractive index and generates a transient grating. Provided spatial and temporal overlap, a third pulse ("probe") can diffract from this grating resulting in a fourth pulse (four wave mixing). If all three input pulses are derived from one single pulse $E(t)$, this process is again a self-referenced gating process with the transient grating acting as the gate $G(t) \propto |E(t)|^2$. Measuring the spectrum of the diffracted pulse as a function of delay between grating and probe, consequently yields a spectrogram of the form

$$S(\Delta t, \omega) \propto \left| \int E(t) \, |E(t - \Delta t)|^2 \, e^{i\omega t} \, dt \right|^2 . \qquad (2.95)$$

Experimentally, the splitting of the input beam into three sub-beams can be achieved by using a mask as shown in Fig. 2.20. Focusing these beams into a thin fused silica substrate that represents the $\chi^{(3)}$ medium yields the required high intensities. For this focusing a spherical bi-mirror is used to allow a variable delay between the two beams generating the grating and the third beam that gets diffracted from it. The

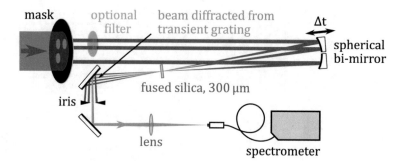

Fig. 2.20 Sketch of the home-built TG-FROG setup

newly created diffracted beam is separated from the others with an iris and sent to a spectrometer.

Since the gate function in the TG-FROG method is proportional to the square of the input field, side wings of uncompressed pulses are partially suppressed which often results in intuitively understandable FROG traces similar to XFROG traces. This useful property can be pushed even further by limiting the bandwidth of the gate pulse after the mask with a bandpass filter. By this means, the effect of higher order dispersion terms on the gate pulse is reduced, making the device de facto a "TG-XFROG".

Another advantage of the TG-FROG principle is that owing to the four-wave-mixing nature of the gating process, the spectral range of the generated signal (i.e. the diffracted pulse) is identical to the range of the input pulses. This allows to characterize pulses with wavelengths down to the UV region, where SHG-FROG-type methods usually fail due to absorption of the generated second harmonic.

The downsides of the implemented TG-FROG are the higher requirements regarding input pulse energy (about $10\,\mu J$ minimum versus $<1\,\mu J$ for the SHG-FROG setup) and the fact that it does not allow single-shot measurements (mechanical scanning device).

2.6 Summary

In summary, in this chapter the theoretical and experimental tools for the description and characterization of ultrashort, intense light pulses have been presented. Furthermore, the most common nonlinear optical processes have been introduced that can be observed and triggered by propagating such pulses in dielectric media. In the following chapters we will refer to these explanations to understand the conducted experiments.

References

1. R.W. Boyd, *Nonlinear Optics* (Academic Press, 2008)
2. R.L. Sutherland, *Handbook of Nonlinear Optics* (Marcel Dekker, 2003)
3. R. Trebino, *Frequency-Resolved Optical Gating: The Measurement of Ultrashort Laser Pulses* (Springer US, Boston, MA, 2000). https://doi.org/10.1007/978-1-4615-1181-6_3
4. N. Karpowicz, V. Yakovlev, Computational methods in photonics, lecture given at Ludwig-Maximilians-Universität München in 2013/2014
5. T. Brabec, F. Krausz, Intense few-cycle laser fields: Frontiers of nonlinear optics. Rev. Mod. Phys. **72**, 545–591 (2000). https://doi.org/10.1103/RevModPhys.72.545
6. T. Brabec, F. Krausz, Nonlinear optical pulse propagation in the single-cycle regime. Phys. Rev. Lett. **78**, 3282, (1997). https://doi.org/10.1103/PhysRevLett.78.3282
7. H. Telle, G. Steinmeyer, A. Dunlop, J. Stenger, D. Sutter, U. Keller, Carrier-envelope offset phase control: a novel concept for absolute optical frequency measurement and ultrashort pulse generation. Appl. Phys. B **69**, 327–332, (1999). https://doi.org/10.1007/s003400050813
8. F. TrAger (ed.), *Springer Hand book of Lasers and Optics* (Springer, Berlin, Heidelberg, 2012). https://doi.org/10.1007/978-3-642-19409-2
9. P.E. Ciddor, Refractive index of air: new equations for the visible and near infrared, Appl. Opt. **35**, 1566–1573 (1996). https://doi.org/10.1364/AO.35.001566
10. H.H. Li, Refractive index of alkaline earth halides and its wavelength and temperature derivatives. J. Phys. Chem. Ref. Data **9**, 161–290 (1980). https://doi.org/10.1063/1.555616
11. D. Zhang, Y. Kong, J.-Y. Zhang, Optical parametric properties of 532-nm-pumped beta-barium-borate near the infrared absorption edge. Opt. Commun. **184**, 485–491 (2000). https://doi.org/10.1016/S0030-4018(00)00968-8
12. M. Daimon, A. Masumura, High-accuracy measurements of the refractive index and its temperature coefficient of calcium fluoride in a wide wavelength range from 138 to 2326 nm. Appl. Opt. **41**, 5275–5281 (2002). https://doi.org/10.1364/AO.41.005275
13. I.H. Malitson, Interspecimen comparison of the refractive index of fused silica. J. Opt. Soc. Am. **55** 1205–1209 (1965). https://doi.org/10.1364/JOSA.55.001205
14. K. Kato, Temperature-tuned 90° phase-matching properties of LiB3O5. IEEE J. Quantum. Electron. **30**, 2950–2952 (1994). https://doi.org/10.1109/3.362711
15. SCHOTT, OpticalGlass-CollectionDatasheets, http://www.us.schott.com/. Accessed 14 Dec 2016
16. D. Eimerl, L. Davis, S. Velsko, E.K. Graham, A. Zalkin, Optical, mechanical, and thermal properties of barium borate. J. Appl. Phys. **62**, 1968–1983 (1987). https://doi.org/10.1063/1.339536
17. D. Strickland, G. Mourou, Compression of amplified chirped optical pulses. Opt. Commun. **56**, 219–221 (1985). https://doi.org/10.1016/0030-4018(85)90120-8
18. J. Moses, C. Manzoni, S.-W. Huang, G. Cerullo, F.X. Kaertner, Temporal optimization of ultrabroadband high-energy OPCPA. Opt. Express **17**, 5540 (2009). https://doi.org/10.1364/OE.17.005540
19. J.-C. Diels, W. Rudolph (ed.), *Ultrashort Laser Pulse Phenomena* (Academic Press, 2006)
20. R. SzipOcs, C. Spielmann, F. Krausz, K. Ferencz, Chirped multilayer coatings for broadband dispersion control in femtosecond lasers. Opt. Lett. **19**, 201 (1994). https://doi.org/10.1364/OL.19.000201
21. V. Pervak, I. Ahmad, M.K. Trubetskov, A.V. Tikhonravov, F. Krausz, Double-angle multilayer mirrors with smooth dispersion characteristics. Opt. Express **17**, 7943 (2009). https://doi.org/10.1364/OE.17.007943
22. A.M. Weiner, D.E. Leaird, J.S. Patel, J.R. Wullert, Programmable femtosecond pulse shaping by use of a multielement liquid-crystal phase modulator. Opt. Lett. **15**, 326 (1990). https://doi.org/10.1364/OL.15.000326
23. P. Tournois, Acousto-optic programmable dispersive filter for adaptive compensation of group delay time dispersionin laser systems. Opt. Commun. **140**, 245–249 (1997). https://doi.org/10.1016/S0030-4018(97)00153-3

24. A. Monmayrant, S. Weber, B. Chatel, A newcomer's guide to ultrashort pulse shaping and characterization. J. Phys. B At. Mol. Opt. Phys. **43**, 103001 (2010). https://doi.org/10.1088/0953-4075/43/10/103001

25. J.E. Midwinter, J. Warner, The effects of phase matching method and of uniaxial crystal symmetry on the polar distribution of second-order non-linear optical polarization. Br. J. Appl. Phys. **16**, 1135–1142 (1965). https://doi.org/10.1088/0508-3443/16/8/312

26. G. Cerullo, S. De Silvestri, Ultrafast optical parametric amplifiers. Rev. Sci. Instrum. **74**, 1–18 (2003). https://doi.org/10.1063/1.1523642

27. I.N. Ross, P. Matousek, G.H.C. New, K. Osvay, Analysis and optimization of optical parametric chirped pulse amplification. J. Opt. Soc. Am. B **19**, 2945–2956 (2002). https://doi.org/10.1364/JOSAB.19.002945

28. S. Demmler, J. Rothhardt, S. HAdrich, J. Bromage, J. Limpert, A. TUnnermann, Control of nonlinear spectral phase induced by ultra-broadband optical parametric amplification. Opt. Lett. **37**, 3933 (2012). https://doi.org/10.1364/OL.37.003933

29. A. Couairon, A. Mysyrowicz, Femtosecond filamentation in transparent media. Phys. Rep. **441**, 47–189 (2007). https://doi.org/10.1016/j.physrep.2006.12.005

30. F. DeMartini, C.H. Townes, T.K. Gustafson, P.L. Kelley, Self-steepening of light pulses. Phys. Rev. **164**, 312–323 (1967). https://doi.org/10.1103/PhysRev.164.312

31. R. del Coso, J. Solis, Relation between nonlinear refractive index and third-order susceptibility in absorbing media. J. Opt. Soc. Am. B **21**, 640 (2004). https://doi.org/10.1364/JOSAB.21.000640

32. N. Matsuda, R. Shimizu, Y. Mitsumori, H. Kosaka, K. Edamatsu, Observation of optical-fibre Kerr nonlinearity at the single-photon level. Nat. Photonics **3**, 95–98 (2009). https://doi.org/10.1038/nphoton.2008.292

33. A. Thai, C. Skrobol, P.K. Bates, G. Arisholm, Z. Major, F. Krausz, S. Karsch, J. Biegert, Simulations of petawatt-class few-cycle optical-parametric chirped-pulse amplification, including nonlinear refractive index effects. Opt. Lett. **35**, 3471–3473 (2010). https://doi.org/10.1364/OL.35.003471

34. N. Minkovski, G.I. Petrov, S.M. Saltiel, O. Albert, J. Etchepare, Nonlinear polarization rotation and orthogonal polarization generation experienced in a single-beam configuration. J. Opt. Soc. Am. B **21**, 1659 (2004). https://doi.org/10.1364/JOSAB.21.001659

35. S. Kourtev, N. Minkovski, L. Canova, A. Jullien, Improved nonlinear cross-polarized wave generation in cubic crystals by optimization of the crystal orientation. J. Opt. Soc. Am. B **26**,1269–1275 (2009). https://doi.org/10.1103/PhysRevLett.84.3582

36. A. Jullien, J.-P. Rousseau, B. Mercier, L. Antonucci, O. Albert, G. ChEriaux, S. Kourtev, N. Minkovski, S. Saltiel, Highly efficient nonlinear filter for femtosecond pulse contrast enhancement and pulse shortening. Opt. Lett. **33**, 2353–2355 (2008). https://doi.org/10.1364/OL.33.002353

37. A. MUnzer, Development of a High-Contrast Frontend for ATLAS Ti:Sa Laser. Diploma thesis, Ludwig-Maximilians- Universität München, 2013

38. H. Fattahi, H.Wang, A. Alismail, G. Arisholm, V. Pervak, A. M. Azzeer, F. Krausz, Near-PHz-bandwidth, phase-stable continua generated from a Yb:YAG thin-disk amplifier. Opt. Express **24**, 24337 (2016). https://doi.org/10.1364/OE.24.024337

39. H. Liebetrau, M. Hornung, A. Seidel, M. Hellwing, A. Kessler, S. Keppler, F. Schorcht, J. Hein, M.C. Kaluza, Ultra-high contrast frontend for high peak power fs-lasers at 1030 nm. Opt. Express **22**, 24776–24786 (2014). https://doi.org/10.1364/OE.22.024776

40. L.P. Ramirez, D. Papadopoulos, M. Hanna, A. Pellegrina, F. Friebel, P. Georges, F. Druon, Compact, simple, and robust cross polarized wave generation source of few-cycle, high-contrast pulses for seeding petawattclass laser systems. J. Opt. Soc. Am. B **30**, 2607 (2013). https://doi.org/10.1364/JOSAB.30.002607

41. A. Buck, K. Schmid, R. Tautz, J. Mikhailova, X. Gu, C.M.S. Sears, D. Herrmann, F. Krausz, Pulse cleaning of few-cycle OPCPA pulses by cross-polarized wave generation, in *Frontiers in Optics 2010/Laser Science*, vol. XXVI, (2010), pp. 8–9. https://doi.org/10.1364/FIO.2010.FMN2

42. R. DeSalvo, M. Sheik-Bahae, A.A. Said, D.J. Hagan, E.W.V. Stryland, Z-scan measurements of the anisotropy of nonlinear refraction and absorption in crystals. Opt. Lett. **18** 194–196 (1993). https://doi.org/10.1364/OL.18.000194

43. V.G. Dmitriev, G.G. Gurzadyan, D.N. Nikogosyan, *Handbook of Nonlinear Optical Crystals*, ed. by A.E. Siegmann (Springer, Berlin, Heidelberg, 1999). https://doi.org/10.1007/978-3-540-46793-9

44. M. Bache, H. Guo, B. Zhou, X. Zeng, The anisotropic Kerr nonlinear refractive index of the beta-barium borate (β-BaB_2O_4) nonlinear crystal. Opt. Mater. Express **3**, 357 (2013). https://doi.org/10.1364/OME.3.000357

45. United Crystals, Properties of LBO Single Crystal, http://unitedcrystals.com/LBOProp.html. Accessed 17 Mar 2017

46. B. Boulanger, J. Zyss, Nonlinear optical properties, in *International Tables for Crystallography*, vol. D, ed. by A. Authier (Wiley, 2013)

47. D. Meschede, *Optik, Licht und Laser* (Vieweg+Teubner Verlag, 2008). https://doi.org/10.1007/978-3-8348-9288-1

48. J.A. Armstrong, N. Bloembergen, J. Ducuing, P.S. Pershan, Interactions between light waves in a nonlinear dielectric. Phys. Rev. **127**, 1918–1939 (1962). https://doi.org/10.1103/PhysRev.127.1918

49. G.M. Gale, M. Cavallari, T.J. Driscoll, F. Hache, Sub-20-fs tunable pulses in the visible from an 82-MHz optical parametric oscillator. Opt. Lett. **20**, 1562 (1995). https://doi.org/10.1364/OL.20.001562

50. G. Arisholm, General numerical methods for simulating Second-Order Nonlinear Interactions in Birefringent Media. J. Opt. Soc. Am. B **14**, 2543–2549 (1997). https://doi.org/10.1364/JOSAB.14.002543

51. C. Skrobol, I. Ahmad, S. Klingebiel, C. Wandt, S.A. Trushin, Z. Major, F. Krausz, S. Karsch, Broadband amplification by picosecond OPCPA in DKDP pumped at 515 nm. Opt. Express **20**, 4619–29 (2012)

52. H. Fattahi, Yb:YAG-pumped, few-cycle optical parametric amplifiers, in *High Energy and Short Pulse Lasers*, ed. by R. Viskup, InTech (2016). https://doi.org/10.5772/61628

53. A.V. Smith, SNLO nonlinear optics code (v.61), http://www.as-photonics.com. Accessed 17 Mar 2017

54. H. Fattahi, *Third-Generation Femtosecond Technology* (Ludwig-Maximilians-Universität München, PhDthesis, 2015)

55. C. Skrobol, *High-Intensity, Picosecond-Pumped, Few-CycleOPCPA* (Ludwig-Maximilians-Universität München, PhDthesis, 2014)

56. T. Lang, A. Harth, J. Matyschok, T. Binhammer, M. Schultze, U. Morgner, Impact of temporal, spatial and cascaded effects on the pulse formation in ultra-broadband parametric amplifiers. Opt. Express **21**, 949–59 (2013)

57. M. Hochbruck, A. Ostermann, Exponential integrators. Acta Numer. **19** (2010). https://doi.org/10.1017/S0962492910000048

58. C. Froehly, A. Lacourt, J. Vienot, Time impulse response and time frequency response of optical pupils. Experimental confirmations and applications. Nouv. Rev. Optique **183**, 183–196 (1973). https://doi.org/10.1088/0335-7368/4/4/301

59. C. Iaconis, I. Walmsley, Spectral phase interferometry for direct electric-field reconstruction of ultrashort optical pulses. Opt. Lett. **23**, 792–794 (1998). https://doi.org/10.1109/CLEO.1998.676573

60. M.E. Anderson, A. Monmayrnat, S.P. Gorza, P. Wasylczk, I.A. Walmsley, SPIDER: A decade of measuring ultrashort pulses. Laser Phys. Lett. **5**, 256–266 (2008). https://doi.org/10.1002/lapl.200710129

61. T. Oksenhendler, S. Coudreau, N. Forget, V. Crozatier, S. Grabielle, R. Herzog, O. Gobert, D. Kaplan, Self-referenced spectral interferometry. Appl. Phys. B Lasers Opt. **99**, 7–12 (2010). https://doi.org/10.1007/s00340-010-3916-y

62. A. Trabattoni, T. Oksenhendler, H. Jousselin, G. Tempea, S. De Silvestri, G. Sansone, F. Calegari, M. Nisoli, Self-referenced spectral interferometry for single-shot measurement of sub-5-fs pulses. Rev. Sci. Instrum. **86** (2015). https://doi.org/10.1063/1.4936289

63. D.J. Kane, Recent progress toward real-time measurement of ultrashort laser pulses. IEEE J. Quantum. Electron. **35**, 421–431 (1999). https://doi.org/10.1109/3.753647
64. R. Trebino, FROG Code, http://frog.gatech.edu/code.html. Accessed 17 Mar 2017
65. R. Trebino, K.W. DeLong, D.N. Fittinghoff, J.N. Sweetser, M.A. KrumbUgel, B.A. Richman, D.J. Kane, Measuring ultrashort laser pulses in the time-frequency domain using frequency-resolved optical gating. Rev. Sci. Instrum. **68**, 3277–3295 (1997). https://doi.org/10.1063/1.1148286
66. S. Akturk, C. D'Amico, A. Mysyrowicz, Measuring ultrashort pulses in the single-cycle regime using frequency-resolved optical gating. J. Opt. Soc. Am. B **25**, A63–A69 (2008). https://doi.org/10.1364/JOSAB.25.000A63
67. J.N. Sweetser, D.N. Fittinghoff, R. Trebino, Transient-grating frequency-resolved optical gating. Opt. Lett. **22**, 519–521 (1997)

Chapter 3
The Petawatt Field Synthesizer System

The PFS system consists of four main sections as shown in the schematic overview in Fig. 3.1:

- The Ti:Sa-based frontend from which all optical pulses of the system are derived.
- The Yb-based amplifier chain that provides high-energy pump pulses for parametric amplification.
- The broadband seed generation section.
- The OPCPA chain where pump and seed pulses are combined.

Framed boxes in the overview indicate elements that—compared to the system status reported in previous works [1–4]—are new in the sense that these elements were greatly modified, have been installed and used for the first time or constitute entirely new developments. In the present chapter, some of these new elements will be described and a summary of the already established components will be given to provide a general understanding of the PFS system. The central topics of this thesis, however, i.e. the development of new seed generation schemes, the OPCPA measurements and the preparations for the PFS upgrade with a third OPCPA stage are discussed in separate chapters.

3.1 Frontend and Initial Seed Generation Approach

For optical synchronization, all pulses of the PFS system are derived from one *Ti:Sa master oscillator* (Rainbow, Femtolasers, 75 MHz, 300 mW) that provides both 6 fs, broadband pulses at 790 nm central wavelength for the OPCPA seed generation as well as narrowband pulses at 1030 nm for the pump chain on a separate output. Compared to the previously implemented version of the oscillator, this additional output (enabled by an optimized intra-cavity dispersion management) constitutes a

© Springer International Publishing AG, part of Springer Nature 2018   59
A. Kessel, *Generation and Parametric Amplification of Few-Cycle Light Pulses at Relativistic Intensities*, Springer Theses,
https://doi.org/10.1007/978-3-319-92843-2_3

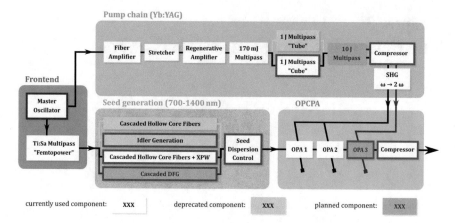

Fig. 3.1 Overview of the PFS system. Framed boxes indicate elements that were newly installed or significantly modified during the course of this work. White boxes indicate the components used during the high-energy OPCPA experiments described in Sect. 5.2. Gray boxes refer to used or developed elements that have been dismissed. Purple boxes indicate elements that are planned for the next extension of the system

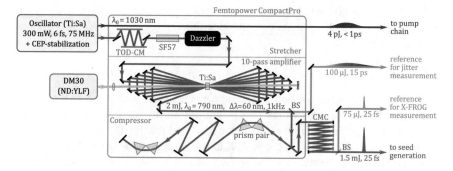

Fig. 3.2 Sketch of the PFS frontend system providing optically synchronized pulses for the pump chain and for seed generation. TOD-CM = third-order-dispersion chirped mirrors, CMC = chirped mirror compressor

valuable improvement as it allows direct seeding of Yb:YAG-based amplifiers. An intermediate photonic crystal fiber stage [2] that had to be used in the past to shift the central frequency of the broadband oscillator pulses into the 1030 nm band therefore became obsolete. The oscillator is furthermore equipped with a feed-forward CEP-stabilization system (CEP4 SEED, Femtolasers) that corrects for fast and slow drifts inside the cavity and thus provides CEP-stable pulses.

To obtain high-energy pulses for the generation of the OPCPA seed, the oscillator output is amplified in a CPA scheme as shown in Fig. 3.2: the pulses are first temporally stretched by TOD chirped mirrors, a 40 mm SF57 substrate and the tellurium dioxide crystal of an acousto-optic modulator (Dazzler, Fastlite) which is used for shaping the spectral amplitude and phase of the pulses. The chirped pulses are then amplified in a 10-pass Ti:Sapphire amplifier (Femtopower CompactPro, Femtolasers) from few-nJ to 2 mJ-pulse energy at 1 kHz repetition rate. A small fraction of the amplified stretched pulses is coupled out by a beam splitter and is later used for synchronization between OPCPA seed and pump beam line (see Sect. 3.3.3). The main beam is sent through a hybrid compressor consisting of a pair of double prisms for pre-compression and an array of chirped mirrors (CM) for final compression to a pulse duration of about 25 fs. This combination of compression techniques proved to be necessary as the exclusive use of prisms resulted in SPM effects in the last prism (for details see [2]). From the compressed beam again a fraction is split off and used as a gate for XFROG measurements. The remaining energy of 1.5 mJ is further used for seed generation.

The initial approach to generate the OPCPA seed pulses in the spectral range from 700 to 1400 nm using the output of the Femtopower was developed by Izhar Ahmad and Sergei Trushin and is described in [5]. It consists of two cascaded hollow core fibers (HCF) and applies the widespread HCF compression technique to produce few-cycle or even sub-cycle pulses from a multi-cycle input [6–11]. This technique is also termed "supercontinuum generation" referring to the spectral broadening associated with the process. The underlying idea is to confine intense light to a small area (the gas-filled void of a hollow-core fiber) over a long propagation distance of typically 1–2 m to trigger high-order nonlinear effects such as SPM and self-steepening. Similarly, supercontinua can be generated by self-focusing and filamentation in bulk material, however, the eventual break up of the filament at high input power limits the available output energy for this method typically to the order of nanojoule to microjoule [12]. HCF broadening on the other hand has been demonstrated with pulse energies of up to several millijoule [13, 14].

The setup of the initial seed generation scheme is shown in Fig. 3.3: The compressed Femtopower output is focused by a lens ($f = 2$ m) into a first HCF (300 μm core diameter, 1 m length) filled with Neon gas. Since proper coupling of the beam into the fiber is crucial for a stable operation, a commercial beam stabilization system (Aligna4D, TEM Messtechnik) was installed that corrects for beam pointing and long term drifts. By adjusting dispersion with the Dazzler and by optimizing gas pressure for a high fiber throughput and good spectral broadening, an output pulse energy of about 650 μJ at a Neon pressure of 2.3 bar was achieved.

At 600–950 nm, the spectral range of the output pulses (see Fig. 3.4) did not fully cover the desired seed spectrum of 700–1400 nm which is due to the intrinsic blueshift of the HCF compression technique that prevents an efficient broadening into the IR in a single-fiber setup [2]. Hence, the pulses were re-compressed in an intermediate chirped mirror compressor to ~5 fs and sent into a second HCF (250 μm diameter, 1 m length, 3.3 bar of Neon) yielding a supercontinuum spanning 300–1400 nm with a total energy of 200 μJ (see Fig. 3.4). By cutting the spectrum with an RG-680 filter (SCHOTT) to the spectral range of interest (700–1400 nm), pulses with an energy of ~30 μJ were obtained.

Even though the pulses produced by this method formally match the spectral and energetic requirements to serve as a seed for the PFS OPCPA chain, we will

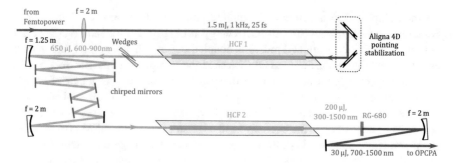

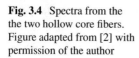

Fig. 3.3 Sketch of the initial seed generation scheme with two cascaded hollow core fibers

Fig. 3.4 Spectra from the
the two hollow core fibers.
Figure adapted from [2] with
permission of the author

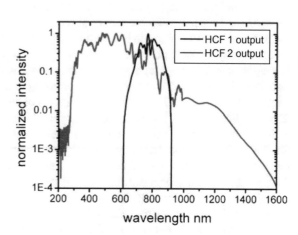

see in Sect. 3.3.1 that the pronounced spectral modulations cause serious distortions
of spectrum and spectral phase during parametric amplification. For this reason,
alternative seed generation schemes have been developed as will be described in
Chap. 4.

3.2   The Pump Chain

The PFS pump chain—as proposed in the original concept [15]—is planned to deliver
sub-picosecond pulses with energies of up to 12 J per pulse in the fundamental beam
at 10 Hz repetition rate. Since a system matching these demanding specifications
is not commercially available, a set of amplifiers as well as a scheme for dispersion control were developed by Izhar Ahmad [2, 5], Sandro Klingebiel [3, 16, 17],
Christoph Wandt [4, 18] and co-workers. In the following we will briefly summarize
the essential ideas and setups and discuss the modifications we performed to further
improve the system.

The pump amplifier chain is based on ***ytterbium-doped yttrium aluminum garnet (Yb:YAG)*** as a gain material which was at the time of design of the PFS system a relatively new material but has become very popular [19–22] for several reasons:

- Despite being a quasi-three-level system Yb:YAG can be operated at room temperature which allows a water- based cooling scheme.
- Compared to the commonly used Nd:YAG (neodymium-doped YAG) crystal, the emission bandwidth of Yb:YAG is significantly broader and hence supports the compression of amplified pulses to shorter pulse durations.
- The absorption maximum at 940 nm [23] allows direct pumping with laser diodes, an advantage over Ti:Sapphire systems where the absorption maximum around 500 nm requires pumping by frequency-doubled Nd-based lasers (which in turn are pumped by flash lamps or diodes). Hence, high-average-power or high-energy OPCPA systems like PFS that employ Yb:YAG benefit from a higher wall-plug efficiency and thus from a smaller energetic, spatial and financial footprint.
- Emitting light at 1030 nm, Yb:YAG has a small quantum defect which implies a low heat aggregation in the crystal [24].

In Fig. 3.5 the main elements of the PFS pump chain are displayed. The 1030 nm output pulses from the oscillator are amplified in two CW-pumped Yb-doped fiber

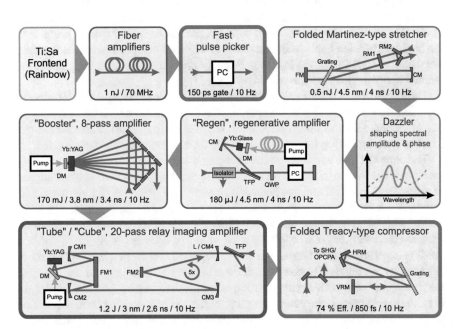

Fig. 3.5 Schematic illustration of the individual elements of the PFS pump chain. Boxes with a thick frame refer to new or modified elements compared to the status reported in [1]. PC = Pockels cell, FM = folding mirror, RM = roof mirror, CM = curved mirror, DM = dichroic mirror, TFP = thin film polarizer, QWP = quarter wave plate, L = lens, VRM = vertical roof mirror, HRM = horizontal roof mirror (Modified version of an original sketch by Mathias Krüger)

amplifiers (developed by the Institute for Applied Physics, Jena, Germany) from 4 pJ to about 1 nJ. To counteract gain depletion of subsequent amplifiers due to the strong amplified spontaneous emission (ASE) background generated in the fiber amplifiers, a fast-switching Pockels cell (Leysop Ltd.) has been installed. By gating the pulses with a ∼150 ps time window, the ASE outside this window is suppressed and its further amplification in subsequent amplifiers is prevented. The benefits of this measure will be further discussed in Sect. 5.2.1.

For a safe amplification of the required high-energy pump pulses the integration of the CPA concept is obligatory. The dispersion control in the pump chain is realized by a matched grating stretcher/compressor. The *stretcher* is a Martinez-type setup in a doubly-folded and reflective variant to reduce the length of the assembly [3, 25]. At a grating line density of 1740 lines/mm, an incident angle of $60°$ and a virtual grating separation of about 6.2 m, it introduces a GDD of $4.9 \times 10^8 \, \text{fs}^2$ and a TOD of $-8.6 \times 10^9 \, \text{fs}^3$ and stretches the pulses to about 4 ns.

A *Dazzler* (Fastlite) allows to adjust higher dispersion orders by controlling the spectral phase of the transmitted pulses and is used for optimization of compression after the CPA chain. Additionally, the spectral amplitude is shaped in order to counteract gain narrowing in the following amplifier stages. By this means, a broad bandwidth and thus a short Fourier-limited pulse duration is maintained.

In a *regenerative amplifier ("Regen")* the shaped pulses are amplified to 180 μJ in about 110 passes at a roundtrip time of 7.4 ns. Due to the large gain of the order of 10^6 this amplifier has been set up with Yb:glass as the gain medium because of the broader emission bandwidth compared to Yb:YAG that prevents excessive gain narrowing. Saturation in the last roundtrips ensures that despite the large gain the output energy fluctuations are small (less than 1%).

The Regen output is further amplified in a half-bow-tie *eight-pass Yb:YAG amplifier ("Booster")* to 170 mJ. To compensate the deterioration of the beam quality in this amplifier, a spatial filter (pinhole, about 75% transmission) is installed in the focus of the magnifying telescope that adapts the beam size for the next stage.

The currently last amplifier of the PFS pump chain is a *20-pass relay imaging amplifier*. The first version of this amplifier was capable of delivering pulse energies of more than 1 J at 2 Hz and has been described in [18]. Because of high intensities at the intermediate foci, the whole setup had to be located in vacuum with a DN-320 tube as a housing, hence the nickname *"Tube"*. While the concept was successfully demonstrated and the Tube could be used at output energies of up to 1 J for first OPCPA experiments (see Chap. 5), a stable long term operation was not possible: at high energy we repeatedly observed damages on the Yb:YAG crystal and imaging mirrors with a corresponding degradation of the beam profile as shown exemplary in Fig. 3.6. In this regard, the compact design of the vacuum housing turned out to be problematic as it strongly hampered the replacement and subsequent optimization of optical components.

Fig. 3.6 Degradated Tube beam profile at 1 J output energy measured at the position of the SHG crystal. Cross sections at different points are shown. Gray contours correspond to median values along the respective dimension

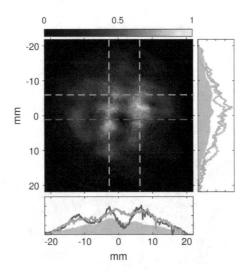

As a consequence, an improved version of the 1 J amplifier was designed and set up by Mathias Krüger and Andreas Münzer. It includes several changes compared to the Tube with the most obvious one being a re-designed vacuum vessel consisting of connected cuboids for convenient access to the optical components, earning it the nickname *"Cube"*. By increasing the beam size from ∼7 to ∼12 mm and by implementing a stronger pump diode (up to 40 J pulse energy, 10.5 J used), an output energy of up to 2 J could be demonstrated. As a reliable and safe point for daily operation, however, a pulse energy of 1.2 J was chosen. Owing to saturation, a good long-term stability of 0.6% rms was achieved. Furthermore, the more elaborate cooling system of the Cube allowed to raise the repetition rate from 2 Hz of the Tube to the full 10 Hz.

The pulses from the last amplifier are finally magnified with an imaging telescope and sent to a folded Treacy-type grating *compressor* that is matched to the stretcher [3, 26]. Compared to prior experiments described in [1], the compressor was re-built in vacuum (Fig. 3.7 shows a sketch of the setup and a photo of the assembly) and connected to the OPCPA system which will be described in the next section. By compressing the pump pulses directly in vacuum, the propagation of compressed pulses in air and the transmission through vacuum windows is avoided and hence the overall B-integral is reduced. This modification allows to use the full ≥1 J output of the last pump amplifier. After optimal alignment of the compressor and adjustment of the spectral phase with the Dazzler, a pulse duration of about 850 fs was achieved—comparable to the previously reported results with the compressor setup in air [1].

(a)

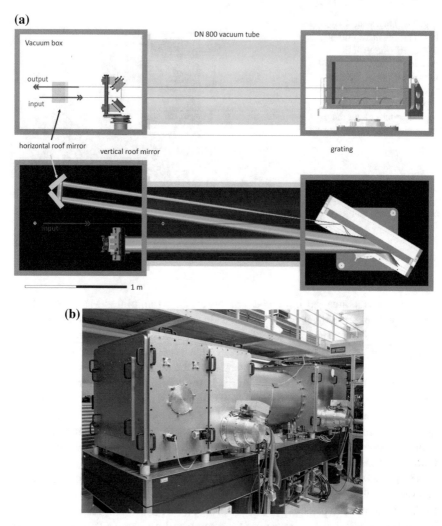

Fig. 3.7 **a** Sketch of the vacuum setup of the folded grating compressor of the PFS pump chain (taken from [3] with the permission of the author). **b** Photo of the assembly in the lab

3.3 The OPCPA System

3.3.1 Previous Experiments

In early experiments performed by Christoph Skrobol and colleagues, two OPCPA stages were set up in air to test the concept of picosecond-pumped parametric amplification [1]. During these measurements, the originally planned DKDP crystals were replaced by LBO in both OPA stages as the latter crystal type showed a better

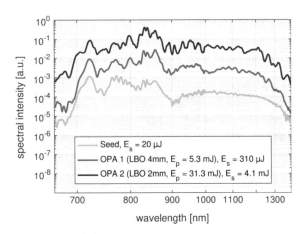

Fig. 3.8 OPCPA measurements with a proof-of-principle setup in air using the initial seed generation scheme consisting of two cascaded HCFs. The spectral modulations present in the seed are enhanced by amplification (note the logarithmic scale). Data taken from [1] with the permission of the author

performance in terms of bandwidth and gain. This replacement was possible due to the comparably small beam sizes of few millimeters and the availability of LBO crystals at such apertures. For the seed pulses the output of the already described cascaded HCFs scheme was used. At the time of measurement, the Booster was the last amplifier in the pump chain providing frequency-doubled pump pulses at 515 nm with energies of 5.3 and 31.3 mJ for the first and second OPCPA stage respectively. The results shown in Fig. 3.8 in general demonstrated the feasibility of broadband amplification but revealed also a serious issue: in contrast to earlier expectations, during amplification the spectral modulations in the signal pulses were not smoothed by saturation but even enhanced. This behavior—that was reproduced later in simulations—can be explained by the influence of the OPA phase: since the signal pulses are chirped, the spectral intensity variations are also present in the time domain and distort—due to the intensity dependent OPA phase—during amplification the spectral phase of the signal pulses. These distortions in turn enhance the amplification in the respective temporal/spectral regions which distorts the phase even more. As a consequence of this positive-feedback process, the amplified pulses eventually feature both a modulated spectrum and a distorted spectral phase and therefore cannot be compressed for subsequent experiments. Furthermore, even in the case of perfect compression the spectral modulations result in temporal pedestals and satellite pulses that contain a significant fraction of the total energy and hence reduce the intensity of the main peak. These observations were the motivation for the development of alternative seed generation schemes that will be discussed in Chap. 4.

Furthermore, the experiments revealed that operating the OPA stages in air limits the usable pump energy to few tens of millijoule due to self-focussing. Hence, an OPCPA vacuum system was designed and constructed by Skrobol et al. [1] which was—largely unchanged—also used for the OPCPA experiments in this work and is described in the following.

3.3.2   The Current System

The current OPCPA vacuum system is shown in Fig. 3.9 as a 3D model (top) and as a schematic layout (bottom). Starting from the compressor exit, the fundamental pump beam is slightly demagnified and spatially filtered with a telescope that images the compressor output onto a DKDP crystal (Type II, $\vartheta = 54.4°$, 4 mm) for second-harmonic generation. A half-wave plate (HWP) in front of the crystal rotates the fundamental polarization to 45°, yielding a second-harmonic polarization parallel to the table. A beamsplitter divides the frequency-doubled beam into a 6% portion for the first and a 94% portion for the second OPCPA stage. While the pump beam for the first stage makes a detour in air for timing adjustment, the pump for the second stage stays in vacuum to minimize B-integral. The residual fundamental is transmitted by the beamsplitter and dumped in air.

Two HWPs are used to independently control the polarization and therefore the effective pump energy for both OPA stages. The pump beams are again imaged by telescopes from the SHG crystal onto the respective OPA crystal. In combination with the above described cascade of telescopes, this ensures that the output of the last amplifier of the pump chain is imaged through the whole system (including compressor) onto the OPA crystals. In this way, the super-gaussian beam profile of the Tube/Cube can—to some extent—be maintained and a deterioration due to a propagation of wavefront errors is suppressed.

The chirped seed pulses (for dispersion management see Sect. 5.2.2) are coupled into the vacuum system from the opposite side and are amplified in the first OPA stage. A retro-reflector mounted on a motorized delay stage is used to adjust the timing for the second OPA stage. With an imaging telescope the beam size of the signal is increased between the stages by a factor of four to match the size of the pump beam at the second stage. After the second stage a rotatable mirror either sends the amplified signal beam to a diagnostic setup in air for characterization of energy and spectrum or guides it to a chamber on a separate table (see the 3D model in the top panel of Fig. 3.9) where the chirped mirrors for compression are located. From there the beam is further transported to the target area (not shown in the model).

3.3.3   Timing Jitter

Even though pump and seed pulses are derived from the same master oscillator, the long and very different optical paths lead to a timing jitter as well as to a slow temporal drift between the pulses. To measure and compensate this effect that eventually has a negative influence on OPCPA stability, a cross-correlation technique [3] is used where the leakage of the pump pulses through a high-reflective mirror (see Fig. 3.9) is combined with an uncompressed part of the Femtopower output (see Fig. 3.2) in a nonlinear crystal as shown in Fig. 3.10 to generate sum-frequency pulses. Due to the chirp in the Femtopower pulses, a relative delay of the pump pulses results in a shift of

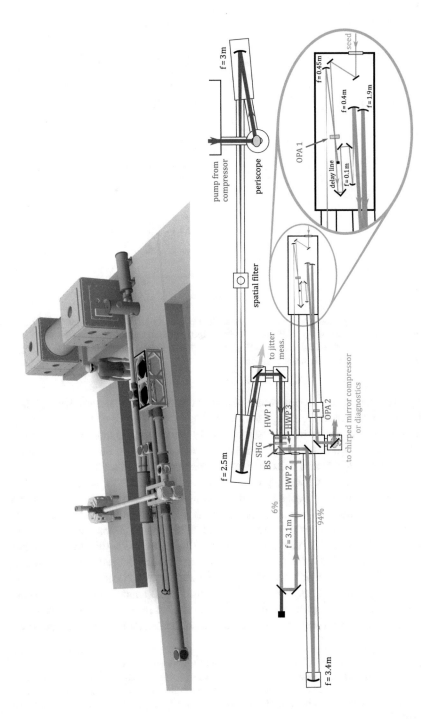

Fig. 3.9 3D model and schematic layout of the PFS OPCPA vacuum system. BS = beamsplitter, HWP = half-wave plate. For details see main text

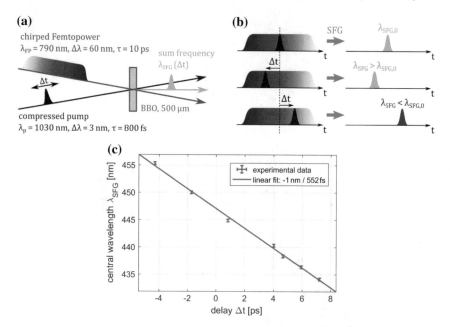

Fig. 3.10 Cross-correlation technique to measure the timing jitter between pump and signal/frontend. **a** Schematic setup. **b** Illustration of how the temporal delay between pulses affects the sum frequency. **c** Calibration measurement to determine the relation between the central wavelength of the sum frequency and the pump pulse delay

the sum frequency. After a calibration, this shift can be used to determine the relative timing for every shot. By feeding the measured delay to a stepper-motor-driven delay stage in the pump chain, both pulses can be synchronized. Assuming furthermore that there is only a constant relative delay between the chirped Femtopower output and the actual OPCPA seed pulses (in good agreement with experimental tests), this method eventually synchronizes pump and seed in the OPCPA crystals.

Initially, the timing jitter measured by Klingebiel et al. in preparation for the first OPA experiments was unacceptably high at 400 fs rms and considerable efforts were made to minimize this value [3]. By theoretical considerations it was suggested that beam pointing due to vibrations and air turbulences inside the pump stretcher and compressor could be the main source of fluctuations.[1] Indeed the construction of a mechanically more stable version of the stretcher setup inside an air-tight box resulted in an improved jitter of ∼100 fs rms, thus small enough for OPCPA experiments but still unsatisfactorily high. Therefore, a great improvement was expected from the rebuild of the compressor in vacuum as described in Sect. 3.2. However,

---

[1]Note, that at the same time a direct contribution of beam pointing via geometrical elongation of the optical path outside the stretcher and compressor was excluded. It can be shown, however, that due to the non-collinear geometry of the diagnostic setup the calculated pointing-induced jitter of 17 as [16] underestimates the contribution by about two orders of magnitude. For details see Appendix A.3.

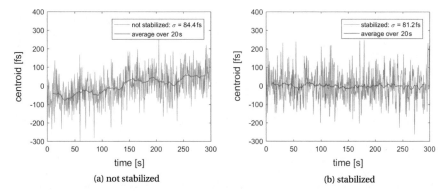

(a) not stabilized   (b) stabilized

Fig. 3.11 **a** Measurement of timing jitter with the cross-correlation technique (see Fig. 3.10) when the stabilization system was switched off. For the measurement the compressed output pulses from the Tube at 2 Hz were used. The pump stretcher was located in an air-tight box and the compressor in vacuum. **b** Measurement under same conditions but with the stabilization system switched on. For stabilization, the average of the last three shots was used (averaging over only two shots or working on a single-shot basis did not result in smaller jitter values)

measurements performed after completion of the rebuild revealed only a slight decrease to about 80 fs rms (see Fig. 3.11). A comparison of Fig. 3.11a, b shows that a temporal drift on the time scale of few seconds can be compensated by the stabilization system but the shot-to-shot fluctuations that dominate the overall jitter can not.

In general, the stabilization is limited to the Nyquist frequency, i.e. half of the acquisition frequency which is given by the repetition rate of the pump. In this respect the change from Tube (2 Hz) to Cube (10 Hz) should have a positive effect which we, however, did not see: in fact the timing jitter even slightly increased when switching to the Cube, indicating on the one hand a strong influence of high-frequency fluctuations and on the other hand that the last pump amplifier might be one of the sources of the remaining jitter. Further investigations are planned for the near future, the achieved temporal stability that corresponds to about 10% of the pump pulse duration, however, allowed a reasonable operation of the OPCPA stages.

References

1. C. Skrobol, High-Intensity, Picosecond-Pumped, Few-CycleOPCPA, PhD thesis, Ludwig-Maximilians-Universität München, 2014
2. I. Ahmad, Development of an optically synchronized seed source for a high-power few-cycle OPCPA system, PhD thesis, Ludwig-Maximilians-Universität München, 2011
3. S. Klingebiel, Picosecond PumpDispersionManagement and Jitter Stabilization in a Petawatt-Scale Few-Cycle OPCPA System, PhD thesis, Ludwig-Maximilians-Universität München, 2013

4. C. Wandt, Development of a Joule-class Yb:YAG amplifier and its implementation in a CPA system generating 1 TWpulses, PhD thesis, Ludwig-Maximilians-Universität München, 2014
5. I. Ahmad, S.A. Trushin, Z. Major, C. Wandt, S. Klingebiel, T. J. Wang, V. Pervak, A. Popp, M. Siebold, F. Krausz, S. Karsch, Frontend light source for short-pulse pumped OPCPA system. Appl. Phys. B: Lasers Opt. **97**, 529–536 (2009). https://doi.org/10.1007/s00340-009-3599-4
6. M. Nisoli, S. De Silvestri, O. Svelto, R. SzipOcs, K. Ferencz, C. Spielmann, S. Sartania, F. Krausz, Compression of high-energy laser pulses below 5 fs. Opt. Lett. **22**, 522–4 (1997)
7. M. Nisoli, S. DeSilvestri, O. Svelto, Generation of high energy 10 fs pulses by a new pulse compression technique. Appl. Phys. Lett. **68**, 2793–2795 (1996). https://doi.org/10.1063/1.116609
8. A. Wirth, M.T. Hassan, I. Grguras, J. Gagnon, A. Moulet, T.T. Luu, S. Pabst, R. Santra, Z.A. Alahmed, A.M. Azzeer, V. S. Yakovlev, V. Pervak, F. Krausz, E. Goulielmakis, Synthesized light transients. Science **334**, 195–200 (2011). https://doi.org/10.1126/science.1210268
9. C. Vozzi, M. Nisoli, G. Sansone, S. Stagira, S. De Silvestri, Optimal spectral broadening in hollow-fiber compressor systems. Appl. Phys. B **80**, 285–289 (2004). https://doi.org/10.1007/s00340-004-1721-1
10. T. Nagy, M. Forster, P. Simon, Flexible hollow fiber for pulse compressors. Appl. Opt. **47**, 3264–8 (2008)
11. B. Schenkel, J. Biegert, U. Keller, C. Vozzi, M. Nisoli, G. Sansone, S. Stagira, S. De Silvestri, O. Svelto, Generation of 3.8-fs pulses from adaptive compression of a cascaded hollow fiber supercontinuum. Opt. Lett. **28**, 1987–9 (2003)
12. M. Bradler, P. Baum, E. Riedle, Femtosecond continuum generation in bulk laser host materials with sub- ?J pump pulses. Appl. Phys. B **97**, 561–574 (2009). https://doi.org/10.1007/s00340-009-3699-1
13. F. Böhle, M. Kretschmar, A. Jullien, M. Kovacs, M. Miranda, R. Romero, H. Crespo, P. Simon, R. Lopez- Martens, T. Nagy, Generation of 3-mJ, 4-fs CEP-stable pulses by long stretched flexible hollow fibers, Research in Optical Sciences: Postdeadline Papers, HW5C.2 (2014). https://doi.org/10.1364/HILAS.2014.HW5C.2
14. A. Suda, M. Hatayama, K. Nagasaka, K. Midorikawa, Generation of sub-10-fs, 5-mJ-optical pulses using a hollow fiber with a pressure gradient. Appl. Phys. Lett. **86**, 111116 (2005). https://doi.org/10.1063/1.1883706
15. Z. Major, S.A. Trushin, I. Ahmad, M. Siebold, C. Wandt, S. Klingebiel, T.-J. Wang, J.A. FÜlÖp, A. Henig, S. Kruber, R. Weingartner, A. Popp, J. Osterhoff, R. HOrlein, J. Hein, V. Pervak, A. Apolonski, F. Krausz, S. Karsch, Basic concepts and current status of the petawatt field synthesizer-a new approach to ultrahigh field generation. Rev. Laser Eng. **37**, 431–436 (2009). https://doi.org/10.2184/lsj.37.431
16. S. Klingebiel, I. Ahmad, C. Wandt, C. Skrobol, S.A. Trushin, Z. Major, F. Krausz, S. Karsch, Experimental and theoretical investigation of timing jitter inside a stretcher-compressor setup. Opt. Express **20**, 3443–3455 (2012). https://doi.org/10.1364/OE.20.003443
17. S. Klingebiel, C. Wandt, C. Skrobol, I. Ahmad, S.A. Trushin, Z. Major, F. Krausz, S. Karsch, High energy picosecond Yb:YAG CPA system at 10 Hz repetition rate for pumping optical parametric amplifiers. Opt. Express **19**, 421–427 (2011)
18. C. Wandt, S. Klingebiel, S. Keppler, M. Hornung, C. Skrobol, A. Kessel, S.A. Trushin, Z. Major, J. Hein, M.C. Kaluza, F. Krausz, S. Karsch, Development of a Joule-class Yb:YAG amplifier and its implementation in a CPA system generating 1 TW pulses. Laser Photonic Rev. **881**, 875–881 (2014). https://doi.org/10.1002/lpor.201400040
19. M. Siebold, J. Hein, M. Hornung, S. Podleska, M.C. Kaluza, S. Bock, R. Sauerbrey, Diode-pumped lasers for ultra-high peak power. Appl. Phys. B: Lasers Opt. **90**, 431–437 (2008). https://doi.org/10.1007/s00340-007-2907-0
20. T. Metzger, A. Schwarz, C. Y. Teisset, D. Sutter, A. Killi, R. Kienberger, F. Krausz, High-repetition-rate picosecond pump laser based on a Yb:YAG disk amplifier for optical parametric amplification. Opt. Lett. **34**, 2123 (2009). https://doi.org/10.1364/OL.34.002123
21. J. Fischer, A.-C. Heinrich, S. Maier, J. Jungwirth, D. Brida, A. Leitenstorfer, 615 fs pulses with 17 mJ energy generated by an Yb:thin-disk amplifier at 3 kHz repetition rate. Opt. Lett. **41**, 246 (2016). https://doi.org/10.1364/OL.41.000246

22. H. Fattahi, H. Wang, A. Alismail, G. Arisholm, V. Pervak, A.M. Azzeer, F. Krausz, Near-PHz-bandwidth, phase-stable continua generated from a Yb:YAG thin-disk amplifier. Opt. Express **24**, 24337 (2016). https://doi.org/10.1364/OE.24.024337

23. J. Körner, J. Hein, M. Kahle, H. Liebetrau, M. Lenski, M. Kaluza, M. Loeser, M. Siebold, Temperature dependent measurement of absorption and emission cross sections for various Yb3+ doped laser materials, in *Proceedings of SPIE*, vol. 8080 (2011), p. 808003. https://doi.org/10.1117/12.887410

24. P.-H. Haumesser, R. GaumÉ, B. Viana, D. Vivien, Determination of laser parameters of ytterbium-doped oxide crystalline materials. J. Opt. Soc. Am. B **19**, 2365 (2002). https://doi.org/10.1364/JOSAB.19.002365

25. O.E. Martinez, 3000 times grating compressor with positive group velocity dispersion:Application to fiber compensation in 1.3-1.6 μm region. IEEE J. Quantum. Electron. **23**, 59–64 (1987). https://doi.org/10.1109/JQE.1987.1073201

26. M. Lai, S.T. Lai, C. Swinger, Single-grating laser pulse stretcher and compressor. Appl. Opt. **33**, 6985–6987 (1994). https://doi.org/10.1364/AO.33.006985

Chapter 4
Seed Generation Schemes

In the previous chapter it was discussed that the strong spectral modulations in the seed pulses generated by spectral broadening in cascaded HCFs constitute a serious problem for the parametric amplification and temporal compression of these pulses. In the present chapter we will describe our efforts to overcome this problem and discuss the advantages and disadvantages of three alternative seed generation schemes we developed.

4.1 Idler Generation

The idler generation scheme that is described in the following is based on optical parametric amplification with the second harmonic (395 nm) of the Femtopower serving as the pump beam and the broadband (500–950 nm) output of a single HCF as the signal beam. Using a Type I BBO in non-collinear geometry, a broad bandwidth can be phasematched (see Fig. 4.1) and the spectrum of the generated idler is located at 700–1400 nm, therefore potentially serving as a seed for the PFS OPCPA chain.

The scheme has a few attractive properties: Since idler photons are only generated during the simultaneous presence of the short (few tens of femtoseconds) pump and signal pulses at the NOPA-crystal, a high contrast can be expected. Furthermore, the temporal gating effectively suppresses high-frequency modulations in the idler spectrum.

The major challenge of the scheme on the other hand is the compensation of the angular dispersion of the idler that originates from momentum conservation in the non-collinear geometry. A possible setup to accomplish this compensation has been proposed by Shirakawa and Kobayashi [1, 2] and is shown in Fig. 4.2. The idea is to image the dispersed idler with a telescope onto a diffraction grating and to match by this means the angular dispersion of the idler with that of the grating to obtain in the end a nearly collimated beam.

© Springer International Publishing AG, part of Springer Nature 2018

A. Kessel, *Generation and Parametric Amplification of Few-Cycle Light Pulses at Relativistic Intensities*, Springer Theses, https://doi.org/10.1007/978-3-319-92843-2_4

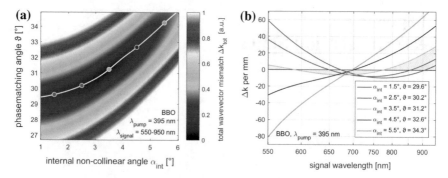

Fig. 4.1 Wavevector mismatch for broadband NOPA in a BBO crystal: **a** total mismatch $\Delta k_{\text{tot}} = \int |\Delta k(\omega_s)| \, d\omega_s$ for different combinations of α_{int} and ϑ. The white line indicates for each α_{int} the optimal ϑ to achieve minimal total mismatch. **b** Wavelength-resolved wavevector mismatch for selected combinations of α_{int} and ϑ

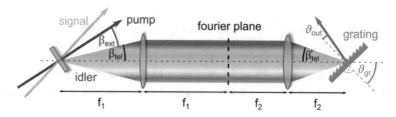

Fig. 4.2 Scheme for compensating the angular chirp of the idler with an imaging telescope and a diffraction grating

4.1.1 Experimental Realization

As a starting point, we used a publication [3] by Wang et al. who built a first test setup of the idler generation scheme at our lab and investigated the scheme theoretically. Despite some conceptual similarities, however, the setup reported here is a complete new development and contains several essential improvements. Additionally, we found some theoretical flaws in [3] that will be addressed later.

The layout of the experimental realization is shown in Fig. 4.3 with the optical path of the individual beams as follows: The output of the Femtopower amplifier is split into two arms containing $\approx 500 \, \mu\text{J}$ (pump arm) and $\approx 330 \, \mu\text{J}$ (seed arm), respectively. A common lens ($f = 2 \, \text{m}$) and beam stabilization system creates a focus of $200 \, \mu\text{m}$ FWHM for both beams.

The **pump arm** is spatially filtered with a pinhole in vacuum and afterwards frequency-doubled in a BBO crystal ($200 \, \mu\text{m}$, $\vartheta = 29.1°$). Because of the long focal length of the focussing lens, the beam divergence at the crystal is small and the effect on phasematching is negligible. After collimation, the fundamental beam is removed by two dielectric mirrors with high-reflectivity at 380–420 nm. The spectrum of the generated second-harmonic is shown in Fig. 4.4a and contains $150 \, \mu\text{J}$ of energy. For

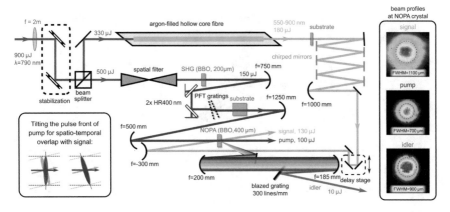

Fig. 4.3 Sketch of the experimental idler generation setup. Find details to the individual components in the text. In insets are shown a symbolical depiction of the introduced pulse front tilt of the pump and the beam profiles of signal, pump and idler beam at the position of the NOPA-crystal

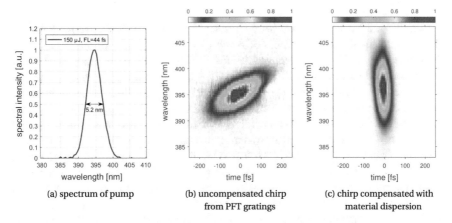

(a) spectrum of pump

(b) uncompensated chirp from PFT gratings

(c) chirp compensated with material dispersion

Fig. 4.4 **a** Spectrum of the pump beam and TG-Frog traces of **b** chirped and **c** compressed pump pulses taken at the position of the NOPA-crystal. The compressed pulse duration was $\sim$45 fs

good spatio-temporal overlap with the seed in the NOPA crystal, the pulse front of the pump was tilted[1] with a pair of lithographic transmission gratings (2500 1/mm line density). To counteract the negative dispersion introduced by the unavoidable gap between the two gratings (making them act effectively as a compressor), 25 mm of fused silica was added to the beam path. By this measure, the pump pulses could be compressed to about the Fourier-limited pulse duration of 45 fs at the NOPA-crystal as measured with a home-built TG-FROG (cf. Fig. 4.4b, c). With 100 µJ of energy and a FWHM beam size of 700 µm at the NOPA crystal, the peak intensity of the pump was 375 GW/cm^2.

---

[1] A detailed description of angular chirp and tilted pulse fronts can be found in Sect. 6.3.

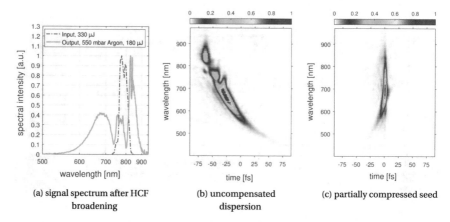

(a) signal spectrum after HCF broadening

(b) uncompensated dispersion

(c) partially compressed seed

Fig. 4.5 a Spectrum of the seed beam and TG-Frog traces of **b** chirped and **c** partially compressed seed pulses taken at the position of the NOPA-crystal

The 330 μJ pulses in the *seed arm* were sent into an HCF (300 μm core diameter, 1 m length) filled with 550 mbar of argon gas. The dispersion of the input pulses was tuned for optimal broadening with the Dazzler inside the Femtopower. Input and broadened spectrum are displayed in Fig. 4.5a. The output pulses have spectral components in the range of 500–950 nm and an energy of 180 μJ. Partial temporal compression was achieved with chirped mirrors and a thin fused silica substrate as shown in Fig. 4.5b, c. Since all relevant seed frequencies lie within the pump duration of 45 fs, no special effort was made to compensate higher order dispersion. The remaining chirp was tuned such that short wavelengths are more stretched, enabling better energy extraction from the pump beam for this spectral range to boost the rather weak long wavelength tail of the generated idler. The seed beam size was adapted with a Galilean-type telescope (to avoid a focus in air), leading to a wavelength-dependent signal beam diameter (larger for longer wavelengths) due to diffraction after the HCF. To support the parametric amplification of short signal wavelengths and hence again the generation of long idler wavelengths, the FWHM diameter of the seed (1100 μm) was set slightly larger than the pump diameter. The temporal overlap with the pump beam was controlled with a delay line.

The *idler* pulses are produced by the interaction of pump and signal in a Type I BBO (500 μm, $\vartheta = 31°$) with an experimentally determined optimal non-collinear angle of 3.7°. As expected, the idler has spectral components ranging from 680 to 1500 m as displayed in Fig. 4.6a with a maximal total energy of 15 μJ and a homogeneous beam profile (cf. Fig. 4.3). The angular chirp was measured by scanning a spectrometer fiber laterally across the dispersed beam behind the NOPA crystal and recording the central wavelength at each position (see Fig. 4.6b).

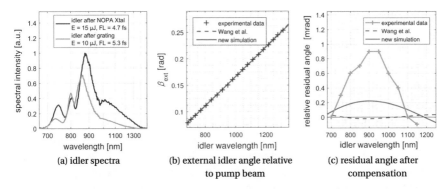

(a) idler spectra

(b) external idler angle relative
to pump beam

(c) residual angle after
compensation

Fig. 4.6 Experimental results for the generated idler: **a** Spectra before and after the telescope-grating-system, **b** wavelength-dependent angle of the idler beam with respect to the pump beam after the NOPA crystal and **c** residual relative angle after the compensation. Simulation results (red curves) were calculated with Eqs. (4.1) and (4.2)

4.1.2   Compensation of the Angular Chirp

To compensate this chirp, the idler beam was imaged with a reflective telescope ($f_1 = 200$ mm and $f_2 = 182.5$ mm) onto a blazed grating (300 lines/mm) using components that had already been suggested and purchased by T. J. Wang. To keep aberrations from the telescope-grating-system as small as possible (NA > 0.1!), the idler beam was folded in the non-dispersed vertical dimension as indicated in Fig. 4.3. At ∼30% losses from the telescope-grating-system, a pulse energy of ∼10 µJ was obtained for the compensated idler. Additionally, the output spectrum was slightly narrowed (cf. Fig. 4.6a) due to the wavelength-dependent diffraction efficiency of the grating.

The residual angle after compensation was measured by focusing the idler with a curved mirror ($f = 500$ mm) onto a CCD. By placing a movable slit into the Fourier plane of the telescope (cf. Fig. 4.2), it is possible to select a narrow spectral band of the idler and detect its focus position on the CCD. Shifting the slit and recording the individual focus positions yields the relative angle for different idler wavelengths. By this method, the telescope and the angle of incidence onto the grating ϑ_{gr} was optimized to achieve a minimal amplitude of the residual angle curve within the spectral range detectable by the CCD ($\lambda \lesssim 1150$ nm). The result is shown in Fig. 4.6c and reveals a residual angular spread of ∼1 mrad, more than a magnitude larger than what has been calculated by Wang.

To understand this discrepancy, we had a closer look at the calculation of $\beta_{ext}(\lambda_i)$ in [3] and it turned out that some crucial aspects had not been considered (for details see Appendix A.4). Including these aspects, we arrive at a more accurate expression for the external angle:

$$\beta_{\text{ext}} = \sin^{-1} \left(\frac{n_{\text{i}} \sin\left(\alpha_{\text{int}}(\lambda_{\text{s}})\right)}{\sqrt{1 + k_{\text{p}}^2/k_{\text{s}}^2 - 2\,k_{\text{p}}/k_{\text{s}} \cos\left(\alpha_{\text{int}}(\lambda_{\text{s}})\right)}} \right) \quad (4.1)$$

where $n_{\text{i}} = n(\lambda_{\text{i}}) = n\left(\frac{\lambda_{\text{s}}\lambda_{\text{p}}}{\lambda_{\text{s}}-\lambda_{\text{p}}}\right)$ is the refractive index of the idler, α_{int} is the internal non-collinear angle between pump and signal and k_{p}, k_{s} are the wavenumbers of pump and signal. Note that as the incidence angle of the signal beam onto the crystal surface depends on the cut angle and is generally not exactly 90°, α_{int} depends due to refraction on the refractive index of the signal n_{s} and therefore on λ_{s}.

Figure 4.7 shows a comparison of both simulations using the experimental parameters. While the differences between Wang's and our calculation seem to be subtle for β_{ext} (cf. Fig. 4.7a), they become obvious for the first (b) and the second-order derivatives (c). Since the goal is to achieve a minimal residual angular dispersion after compensation, these higher orders are important.

Calculating the wavelength-dependent output angle ϑ_{out} of the idler beam after the telescope-grating compensation scheme yields

$$\vartheta_{\text{out}}(\lambda_{\text{i}}) = \sin^{-1}\left[\frac{\lambda_{\text{i}}}{d} - \sin\left(\tan^{-1}\left(\frac{\tan\left(\beta_{\text{tel}}(\lambda_{\text{i}})\right)}{M}\right) + \vartheta_{\text{gr}}\right)\right] \quad (4.2)$$

where d is the line separation of the grating, $M = f_2/f_1$ is the telescope magnification and ϑ_{gr} is the angle between grating surface normal and optical axis of the telescope. For minimal aberrations, the system is set up such that this axis is the angle bisector of the bundle of idler rays. Hence,

$$\beta_{\text{tel}}(\lambda_{\text{i}}) = \beta_{\text{ext}}(\lambda_{\text{i}}) - \beta_{\text{ext}}(\lambda_{\text{i,central}}) \quad (4.3)$$

with the central idler wavelength $\lambda_{\text{i,central}} = 1040\,\text{nm}$.

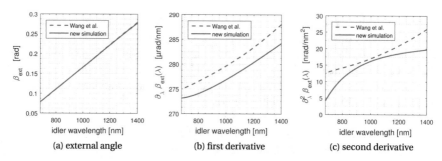

Fig. 4.7 Theoretical prediction for external angle and angular dispersion of the generated idler (red curve). For comparison the curves obtained by Wang et al. [3] are shown

The results of the corrected theoretical predictions for β_{ext} and $\vartheta_{out}(\lambda_i)$ are plotted together with the experimental results in the already displayed Figs. 4.6b, c. While for β_{ext} the precision of the experimental data is not high enough to discriminate between the two theoretical predictions, the measured residual angles are clearly better described by the new simulation than by Wang's prediction. However, there is still a noticeable deviation between experiment and theory.

To estimate with what accuracy the angular chirp has to be compensated, a model suggested in [4] is used that quantifies the effect of angular dispersion on the intensity of a focused and temporally compressed beam with Gaussian spectrum and spatial profile. The derived factor

$$\xi = \sqrt{1 + \left(C_a \frac{\pi}{2 \ln 2} \frac{\Delta\lambda_{FWHM}}{\lambda_0} d_{FWHM} \right)^2} \qquad (4.4)$$

defines the temporal elongation of the pulse as well as the enlargement of the focal spot along the dimension of dispersion. The quantities in the equation are the angular dispersion C_a, the bandwidth $\Delta\lambda_{FWHM}$ of the pulse, the central wavelength λ_0 and the beam size d_{FWHM}. Since both pulse duration and focal spot size are affected by angular dispersion, the reduction of intensity is inversely proportional to ξ^2.

From the measured residual relative angles shown in Fig. 4.6c, one can deduce an angular dispersion of up to $5\,\mu\text{rad/nm}$, which would according to Eq. (4.4) imply an unacceptable peak intensity reduction by ξ^2 of up to 10. We define here $\xi^2 = 1.5$ as an acceptable value, i.e. allow for a 33% reduction of peak intensity. This value is chosen rather generously being aware of the fact that the spectrum of the idler is not Gaussian and the residual angular dispersion is not constant over the full spectral range and therefore the actual impact on peak intensity is less drastic than the nominal value of 33% suggests. To achieve this defined limit, a reduction of the angular dispersion C_a to about $1\,\mu\text{rad/nm}$ would be necessary.

To determine the potential for improvement when choosing another combination of telescope magnification M and grating line density $1/d$, we performed a ***global optimization*** with these free parameters and the grating angle ϑ_{gr}. A perfect compensation of the angular chirp of the idler beam is equivalent to achieving an identical ϑ_{out} for all λ_i. Hence, as a cost function for numerical optimization, we chose the standard deviation of the residual dispersion $\partial_{\lambda_i}\vartheta_{out}(\lambda_i)$ from zero. Figure 4.8a shows the results for the experimentally realized case of $M = 0.9125$ and $d = 1/300$ mm and for the global optimum. As one can see, the simulation suggests that by choosing a larger magnification and a smaller line density, a reduction of the residual angular dispersion by a factor of 10 is possible, at least in theory.

However, a ***stability analysis*** reveals that achieving this optimal compensation is not a trivial task. In Fig. 4.8b the residual angular dispersion for a range of telescope magnifications and grating line densities is plotted where for each combination of M and $1/d$ the angle ϑ_{gr} was optimized to obtain a minimal residual angular dispersion.

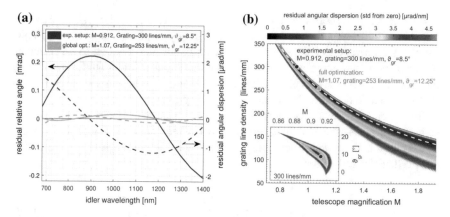

Fig. 4.8 **a** Simulated residual relative angle and angular dispersion of the idler beam after compensation with a telescope and grating. Purple curves correspond to the experimentally realized configuration of telescope and grating where only ϑ_{gr} was used for optimization. Olive curves correspond to the case that all three parameters M, d and ϑ_{gr} were optimized. **b** Residual angular dispersion in a 2D parameter space. For details see text

Apparently, all good combinations of M and grating line density follow a distinct line (white dashed) where the parameter space with acceptable residual dispersion (i.e. pixels with colors black to purple-blue) is narrower when choosing higher grating line densities. Both, the experimental setup as well as the setup suggested by global optimization, are located in this region of high line density. As a consequence, M and ϑ_{gr} have to be adjusted very carefully as can be deduced from the inset of Fig. 4.8b for a grating with 300 lines/mm. A deviation of M by 1% from the optimum—e.g. due to misalignment or a slightly wrong focal length of one of the telescope mirrors—can already make it impossible to achieve an acceptable compensation, even if ϑ_{gr} is afterwards fully optimized. In practice, the global optimizing of M and ϑ_{gr} is a cumbersome process as for each measurement point the grating has to be positioned in the image plane of the telescope, the beam has to be re-aligned onto the CCD after setting the grating angle and the residual angular chirp has to be measured by scanning over the wavelengths.

Furthermore, experience from daily operation as well as theoretical considerations show that any slight change of the non-collinear angle α_{ext} between pump and signal does not only influence the direction of the idler beam but also its angular dispersion. For instance, changing α_{ext} by 200 μrad results in an already critical change of angular dispersion by more than 0.5 μrad/nm for all wavelengths. At an angular magnification of ~3 by the telescopes of pump and signal before the NOPA-crystal, a misalignment of this order of magnitude can easily happen during a measurement day, thus requiring monitoring and a (possibly time-consuming) realignment of the compensation setup. Since the seed generation setup constitutes just one of many elements of the whole PFS system, a less maintenance-prone solution is needed.

4.1.3 Possible Improvements of the Setup

There are a few conceivable options to reduce the residual angular chirp and to improve the stability of the idler generation scheme:

Since the described instability in the experimentally realized setup would apply in the same way for the *globally optimal configuration* (M = 1.072, 253 lines/mm, ϑ_{gr} = 12.25°), there is not much improvement to expect from the realization of this setup. Additionally, it remains unclear if the deviation of the experimental residual angle from the theoretical prediction in Fig. 4.6c is only due to non-perfect optimization or if there is a systematic offset from theory that would also negatively affect the achievable compensation for any other grating-telescope combination.

Having another look at Fig. 4.8b it seems advantageous to work at a *lower line density* and a larger magnification to reduce the sensitivity to misalignment. However, Eq. (4.4) teaches us that to maintain a fixed pulse lengthening ξ, the angular dispersion C_{a} has to scale inversely proportional to the beam diameter d_{FWHM}. Hence, for a magnification of M = 2 the required C_{a} would be 0.5 μrad/nm (instead of 1 μrad/nm) which is already slightly lower than the theoretical limit for this setup.

Another option would be the installation of an *adaptive mirror* as folding mirror in the Fourier plane of the telescope (indicated in Fig. 4.2). The ability to control the angle of the mirror surface at different positions (=wavelengths) allows in principle for a perfect compensation of the residual angular chirp. But besides adding more complexity to the setup, an unfavorable angular-temporal entanglement would be introduced: since an adaptive mirror generates the required angles via local longitudinal shifts of the mirror surface, a wavelength dependent delay is introduced at the same time which adversely affects temporal compression.

It appears that the only viable solution would be to work with *smaller beam profiles* at the NOPA-crystal, i.e. to reduce d_{FWHM} in Eq. (4.4). But since we already worked rather close to the damage threshold intensity of the NOPA-crystal, one would have to simultaneously reduce the total energy, a measure that is not very satisfactory. A way to circumvent this energy reduction could be to create line foci (or strongly elliptical beams) for pump and seed where the minor axis would be in the plane of angular dispersion. By this means the effective d_{FWHM} could be reduced without increasing the intensity above damage threshold. Furthermore, due to the decreased beam size along the non-collinear dimension, the PFT gratings for the pump would not be necessary anymore.

Since the experimental realization of this idea, however, would have required a major rebuild of the setup and since we found a simpler scheme for seed generation in the meantime (see next section), the development of the idler generation scheme was discontinued at that point.

4.2   Cross-Polarized Wave Generation

The idler generation scheme described in the last section can be understood as an attempt to avoid the problems of the original cascaded HCF scheme (cf. Sect. 3.3.1) by implementing an entirely different scheme. In parallel, we followed another, more straightforward route: Since the problems we faced with the cascaded HCFs arose from the strong modulations in the spectrum of the seed pulses, the obvious solution would be to spectrally smooth these pulses in a subsequent step. In practice, this smoothing can be realized by temporal gating of the pulses. Owing to the required ultrashort switching time of few femtoseconds, this gating has to be based on an optical process. XPW generation that has been described mathematically in Sect. 2.2.2 performs exactly this task and experimentally proved to be a reliable method to spectrally broaden and improve the contrast of short [5, 6] and ultrashort [7–9] pulses. Hence, the scheme described in the following was developed using this technique.

The setup we built for temporal gating of the modulated pulses is shown in Fig. 4.9: The output of the second HCF is filtered with a dielectric longpass filter (LP900, Asahi Spectra, $\lambda_{cut} = 900$ nm) and imaged with a spherical mirror (f = 1500 mm) onto a BaF_2 crystal (1 mm, z-cut) that serves as $\chi^{(3)}$-medium for the XPW generation process. In between, a polarizer (Thorlabs, LPVIS) selects the p-polarization from the slightly elliptically polarized output of the HCF and a pair of fused silica wedges and four chirped mirrors (PC503, UltraFast Innovations) are used for temporal compression. One might notice that as the beam is not collimated, the angle of incidence for example at the chirped mirrors slightly changes across the beam profile due to the bent wavefront. However, since the convergence angle is small ($<0.2°$) no significant negative effect is to be expected. After the BaF_2 crystal the beam is collimated for transport to the OPCPA stages. A second polarizer with crossed orientation relative to the first one selects the s-polarized XPW.

The LP900 longpass filter serves three important purposes: first, it protects the subsequent silver mirrors against ultraviolet radiation of the HCF output pulses that would otherwise damage them over time. Second, it shifts the central wavelength of the filtered pulses to about 1050 nm, supporting spectral broadening in the range of interest (700–1400 nm) by SPM and XPW generation. And finally it removes weak but temporally long artifacts at wavelengths below 900 nm that we found are otherwise able to leak through the crossed polarizers. In a prior configuration with

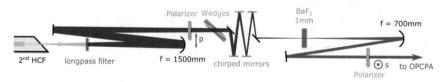

Fig. 4.9 Sketch of the XPW seed generation setup. The output of the second HCF is temporally compressed and focused into a BaF_2 crystal. A crossed polarizer pair selects the generated XPW. For details about the used components see the main text

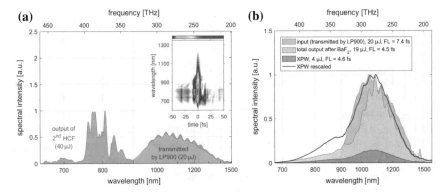

Fig. 4.10 a Output spectrum of the second HCF (wavelengths <650 nm not shown) and transmitted spectrum by the LP900 longpass filter. The inset shows a TG-FROG trace of the compressed pulses at the BaF$_2$ crystal where an RG680 longpass filter was used to allow for the measurement of a broader spectral range. **b** Input and output spectra of the XPW generation process

an RG-1000 filter (SCHOTT) featuring a smooth but very broad cut-off around 1000 nm these spectral component were not entirely removed and experienced large gain during amplification in the OPCPA stages and hence distorted once more the amplified spectrum and phase. Therefore, we decided to use the described dielectric filter with a hard cut-off.

In Fig. 4.10a the modulated output of the second HCF is displayed together with the spectrum of the pulses transmitted by the longpass filter. Note that wavelengths below 650 nm that contain a large fraction of the 150 μJ total energy are not shown here. The FROG trace (see inset) was measured with a home-built TG-FROG at the position of the BaF$_2$ crystal and demonstrates the decent overall compression of the HCF pulses that is required to achieve a high XPW conversion efficiency [10]. The FROG measurement was taken using an RG680 longpass filter instead of the LP900 to show a broader spectral range. Therefore at wavelengths below ∼900 nm the mentioned temporal artifacts are visible that are otherwise cropped by the LP900 filter in the normal configuration.

Figure 4.10b shows the spectra measured in front of and behind the BaF$_2$ crystal. As one can see by comparison of the input and the *total* output spectrum (which contains both polarizations: input and crossed) there is a significant contribution from SPM to spectral broadening towards shorter wavelengths. This is a general phenomenon as efficient XPW generation requires very high intensities (∼15 TW/cm^2 in our setup) and hence other $\chi^{(3)}$—effects such as SPM are inevitably triggered simultaneously [11]. Compared to the total output, however, the XPW spectrum is much less modulated as is expected from the temporal gating of the process. Due to this gating and the decent compression of the input pulses, the pulse duration of the cross-polarized pulse after the BaF$_2$ crystal can be assumed to be close to the Fourier limit of 4.6 fs. FROG measurements of the stretched and re-compressed pulses after the OPCPA stages will be presented in Sect. 5.2.3 and verify the good compressibility.

The measured XPW generation efficiency is at about 20% fairly good considering the predicted theoretical limit for Gaussian pulses of 28% [12]. Compared to the anticipated 20 µJ of seed energy in the original design of the OPCPA stages [13], however, the XPW energy of 4 µJ is considerably lower—a drawback of this scheme. Yet, we will see that saturation of the OPCPA stages to some extent compensates this decrease. On the positive side, the smooth spectrum and spectral phase of the XPW promise a significant improvement of the compressibility of the amplified pulses. This property together with the reliability of the setup made the XPW scheme our standard seed generation scheme for daily operation.

4.3 Cascaded Difference-Frequency Generation

In the last section it was demonstrated that seed generation by two cascaded HCFs and a subsequent XPW stage is a reliable method to generate broadband pulses with a smooth spectral phase. The final seed pulse energy of about 4 µJ, however, reflects an unsatisfyingly low overall seed generation efficiency considering the available energy of 1.5 mJ from the Femtopower. More importantly, simulations show that higher seed pulse energies result in a better extraction of pump energy in the OPCPA stages [13] and hence allow to use thinner OPA crystals for the same output energy. This in turn would on the one hand relax phase-matching conditions leading to broader amplified bandwidths and on the other hand reduce parametric fluorescence as well as parasitic amplification of temporal artifacts.

Therefore, in parallel to the OPCPA experiments described in the next chapter, we continued our search for a robust and efficient seed generation scheme and became aware of a method suggested by Fattahi et al. [14]: they were able to convert a 450–950 nm spectrum to the 1000–2500 nm range (short-wavelength infrared, SWIR) by collinear difference-frequency generation in an appropriately phase-matched BBO crystal. At a reported conversion efficiency of more than 12% this method appeared to be an attractive way to first shift the central frequency of spectrally broadened Ti:Sa pulses to the SWIR region and to broaden the obtained pulses afterwards in another HCF. Because of the typical blue-shift in HCF broadening, in the end a substantial energy fraction can be expected in the range of 700–1400 nm. The fact that the reported central wavelength ($\approx$2000 nm) of the SWIR pulses is located outside this spectral range of interest would have been quite beneficial here as the characteristic spectral and phase distortions caused by SPM are usually confined to a narrow spectral band around the central wavelength. Hence, the combination of DFG and subsequent broadening promised to generate high-energy pulses with a smooth spectral intensity and phase in the range of 700–1400 nm.

Our attempts to reproduce the DFG results in [14], however, were not success-ful as the measured spectrum of the generated pulses peaked around 1000 nm and not as reported above 2000 nm (cf. Fig. 4.13 later in this chapter). Upon a closer analysis, it turned out that the strongly wavelength-dependent spectral sensitivity of the spectrometer (NIRQuest512-2.5, Ocean Optics) shared by our groups had not been taken into account in [14]. Therefore one has to assume that the reported DFG conversion efficiency was overestimated, rendering the planned scheme of DFG and HCF broadening unattractively inefficient.

On the other hand, the output energy and spectrum that was measured in our setup matched the requirements for a PFS seed quite well. This was at first glance surprising since the observed spectrum could not be explained by simple DFG alone. Therefore we investigated the scheme further to understand the underlying physics and to examine its potential usability for our purposes. The results of this investigation are published in [15] and will be summarized in the following.

4.3.1 Experimental Setup and Findings

The experimental setup is shown in Fig. 4.11 and starts with the broadened output of the first HCF of the cascaded HCF seed generation scheme. The gas pressure inside the fiber was set to 3 bar of Neon to provide a bandwidth of 500–950 nm. The pulses are compressed by chirped mirrors (PC70, UltraFast Innovations) and a pair of fused-silica wedges to about 4.7 fs (Fourier limit: 4.1 fs) and are linearly polarized by a thin nano-particle polarizer (Thorlabs, LPVIS). A silver-coated filter wheel is used to fine control the power. The beam is then focused with a spherical mirror into a BBO crystal (Type I, $\vartheta = 20°, \varphi = 90°, d = 500\,\mu m$) and afterwards recollimated by another spherical mirror. A second polarizer is used to select the (relative to the BBO) ordinary polarized components which are then analyzed with a power meter and a set of three calibrated broadband spectrometers (Ocean Optics).

A crucial point for the experiment is the rotation angle ψ of the BBO crystal (the roll-angle in the "yaw-pitch-roll" terminology, compare also Fig. 2.10): If the crystal is oriented such that the input beam is polarized along its extraordinary axis, the output beam is also exclusively extraordinarily polarized (see Fig. 4.12a) and no light passes the ordinary oriented second polarizer. However, if one rotates the BBO by ψ around the optical axis, the input beam splits when it enters the crystal because of birefringence into ordinary and extraordinary components, where the respective

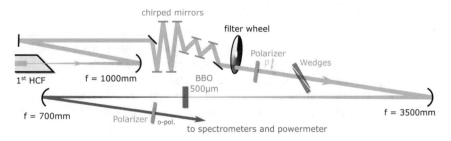

Fig. 4.11 Scheme of the cascaded DFG seed generation setup

(a)   **(b)**

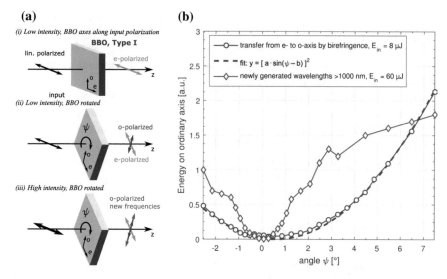

Fig. 4.12 a Sketch of the polarization directions. Linearly polarized input pulses are split into extra-ordinary (green) and ordinary (blue) polarized components depending on the roll-angle ψ of the BBO crystal (i) & (ii). For high intensities, new frequencies are generated on the ordinary axis (iii). **b** Measured energy on the ordinary axis in dependence of ψ. The blue curve shows the $\sin^2$-scaling for weak input pulses ($8\,\mu J$, 500–950 nm) in agreement with Eq. (4.5). The input pulses were additionally temporally stretched with a block of glass to suppress any nonlinear effects. The red curve shows the generation of new wavelengths ($\lambda > 1000$ nm selected by an RG1000 filter) for high intensity input pulses ($E_{in} = 60\,\mu J$). The small x-offset of the red curve with respect to the blue one can be explained by a slight nonlinear polarization rotation in air before the crystal for intense pulses. The non-zero power of the blue data points around the minimum of the curve originates from the imperfect contrast of the used polarizers for $\lambda < 600$ nm

electric field amplitudes A_o and A_e depend on the field amplitude A_{in} of the input as well as on the angle ψ:

$$\text{(a)} \quad A_o = A_{in}\,\sin\psi \qquad\qquad \text{(b)} \quad A_e = A_{in}\,\cos\psi \qquad\qquad (4.5)$$

The energy that is contained in the two polarization directions is by definition proportional to $|A_{o/e}|^2$ and hence proportional to $\sin^2\psi$ and $\cos^2\psi$. This scaling was confirmed experimentally with weak, temporally stretched pulses to suppress any nonlinear contributions (see blue curve in Fig. 4.12b). If one increases the intensity to several TW/cm^2 in the focus ($E_{in} = 60\,\mu J$) and spectrally filters the output beam with a long-pass filter (RG1000), the generation of new wavelengths above 1000 nm can be observed (red curve in Fig. 4.12b). A strong dependence of the energy in this spectral region on ψ is visible as well as the existence of an angle where no new frequencies are generated. In the following experiments, ψ was set to about 5° as this angle proved to give best results in terms of energy and spectral shape of the generated pulses.

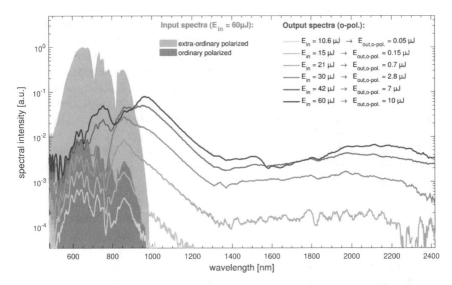

Fig. 4.13 Experimentally generated spectra in a 500 μm BBO for different input energies. Find details about the involved nonlinear processes in the text

The measurement of the spectrum of the o-polarized output pulses reveals a continuum that spans more than two octaves from 500–2400 nm (see Fig. 4.13). This continuum contains spectral bands of different origins:

1. In the region 500–950 nm we first of all find the ordinary-polarized fraction of the input beam ("seed" or "signal") corresponding to the blue curve in Fig. 4.12b. For $\psi = 5°$ its energy equals less than 1% of the total input energy while more than 99% is contained on the e-axis ("pump").

2. By OPA, the long wavelength part (>700 nm) of the signal is amplified by the short wavelength part of the pump, for example:

$$\omega_{p,\lambda=600\,nm}\,(e) + \omega_{s,\lambda=825\,nm}\,(o) \rightarrow 2 \times \omega_{s,\lambda=825\,nm}\,(o) + \omega_{i,\lambda=2200\,nm}\,(o) \quad (4.6)$$

This parametric amplification of the signal boosts the output energy in the range of 700–950 nm significantly.[2] By the same process, the SWIR components above $\sim$1400 nm are generated as the difference frequency ("idler") of pump and signal—represented in Eq. (4.6) by $\omega_{i,\lambda=2200\,nm}$. This is also the interaction that has been described in [14].

3. The most interesting spectral region is that between $\sim$950–1400 nm which contains a large fraction of the total energy. As mentioned before, the generation of photons in this band is at first glance surprising since they cannot be efficiently produced by a single $\chi^{(2)}$—process given the range of the input spectra. A strong influence of

---

[2]It should be pointed out that the broad bandwidth of the interaction would usually (i.e. for a narrowband pump) require a non-collinear geometry for phase matching. Because of the broadband pump employed here, however, this requirement is relaxed permitting the described collinear OPA.

$\chi^{(3)}$—processes on the other hand can be excluded: if XPM by the pump caused a broadening of the signal into the infrared, one should also observe SPM in the pump spectrum (which we did not) as both effects generally appear at a similar intensity level [11]. Additionally, the spectral broadening via XPM would strongly depend on intensity, whereas we can see in Fig. 4.13 that the spectral *shape* between 950 and 1400 nm is hardly changing with increasing input energy. XPW generation does not serve as a proper explanation either since the XPW would not vanish for $\psi = 0$ which was the case in our measurements as already shown in Fig. 4.12b.

Our explanation for the findings is therefore the cascading of two $\chi^{(2)}$—processes, an effect that has been reported before in other experiments [16–18]. An exemplary interaction would be:

$$\text{step 1}: \quad \omega_{p,\lambda=600\,\text{nm}}\,(e) \quad - \quad \omega_{s,\lambda=825\,\text{nm}}\,(o) \rightarrow \quad \omega_{i,\lambda=2200\,\text{nm}}\,(o)$$

$$\text{step 2}: \quad \omega_{p,\lambda=700\,\text{nm}}\,(e) \quad - \quad \omega_{i,\lambda=2200\,\text{nm}}\,(o) \rightarrow \quad \omega_{s,\lambda=1027\,\text{nm}}\,(o)$$

So the idler photons generated in a first DFG process by one pump wavelength (600 nm) interact subsequently with pump photons of a different wavelength (700 nm) in a second process and generate photons in the sought-after spectral region. Both processes are well phasematched by the employed BBO crystal leading to an efficient cascading.

4.3.2 Simulation

To verify our explanation that cascading processes contribute significantly to the generation of the measured output pulses, we performed simulations with the exponential Euler method that was described in Sect. 2.4. Because of the small thickness of the crystal and the long Rayleigh-length of the beam, the plane wave approximation of the simulation is justified and effects such as walk-off, diffraction and self-focusing can be neglected. The effective nonlinear susceptibility tensors $\chi^{(2)\prime}_{lmn}$ and $\chi^{(3)\prime}_{mnpq}$ were obtained by using Eqs. (2.84) and (2.85) and the rotation angle $\psi = 5°$ was accounted for by scaling the amplitudes of the input fields on ordinary (x) and extraordinary (y) axes in accordance with Eq. (4.5). For the spectrum and spectral phase of the input fields the experimental spectra and the retrieved phase of a TG-FROG measurement were used.

As mentioned before, the simulation code contains only one spatial dimension, namely the direction of propagation. However, by segmenting the lateral beam profile into a 2D-grid and assuming that the electric fields within one segment propagate independently of the fields in other segments, a spatially dependent intensity profile can be taken into account (quasi 3D+time). For the simulation we used a Gaussian profile with 380 μm FWHM diameter and circular symmetry in good agreement with a CCD-measurement of the real beam. The profile was divided into 200 × 200 segments and the corresponding intensity values were retrieved. After running the

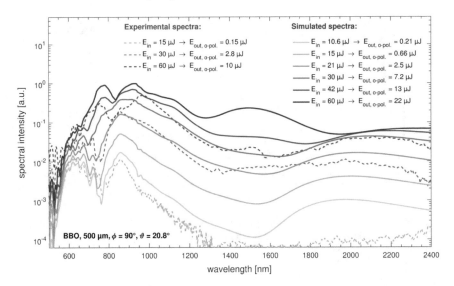

Fig. 4.14 Simulated output spectra of the broadband cascaded nonlinear interaction. For comparison three measured spectra are plotted

simulation for a range of 60 different input intensities to create a lookup-table of generated spectra, each segment can be linked to a spectrum in the lookup-table according to its intensity value. The summed up spectra of all segments yield then the global spectrum.

The results of these calculations—i.e. the generated global spectra—are shown in Fig. 4.14. To allow the comparison with experimental data, the same total energies were used as input parameter. As one can see, the creation of a multi-octave-spanning spectrum is qualitatively well reproduced and especially the efficient generation of new frequency components at 950–1400 nm is confirmed. The simulated spectra in fact extend to about 3000 nm, suggesting that also the experimental spectra might have components up to that wavelength (which we could not measure with the available spectrometers). Including or neglecting third-order effects in the simulation did not affect the results significantly which confirms that cascaded $\chi^{(2)}$—interactions are the dominant factor. Quantitatively, the order of magnitude of the output energy is correctly calculated but a higher conversion efficiency compared to the experimental results is predicted. This might originate from overestimating the peak intensity of the input pulse due to temporal and spatial artifacts from the hollow-core fiber broadening and chirped mirror compression or from imperfections in the BBO crystal.

To visualize the cascading nature of the nonlinear interactions, we additionally ran a simulation with narrow spectral bands but otherwise unchanged parameters (cf. Fig. 4.15). The small bandwidths allow to identify the underlying frequency-mixing processes and the result confirms the contribution of cascaded interactions to the formation of the multi-octave spectrum.

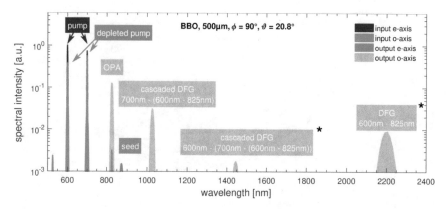

Fig. 4.15 Simulated output of nonlinearly interacting narrowband pulses: two pump wavelengths (600 and 700 nm) on the e-axis and one seed/signal wavelength (825 nm) on the o-axis. The different frequency-mixing products are labeled with the underlying processes. Products marked with an asterisk are passively CEP stable

It should be noted at this point that if CEP stability is of interest for an application, only DFG-products generated by an odd number of cascaded $\chi^{(2)}$—processes are passively phase-stable, while the other products inherit the phase of the input pulses [19]. This is also important if the entire spectral bandwidth is planned to be used since in the case of a non-CEP-stable input beam a variable phase jump at the border ($\sim$1400 nm) between these two spectral regions would prevent full temporal compression of the output pulses. However, simulations show that for actively stabilized systems, CEP fluctuations of $\sim$300 mrad and intensity fluctuations of up to $\pm$20% are tolerable to maintain a stable waveform of the output pulses. Under these practically feasible conditions the achievable pulse duration of 2.6 fs (sub-cycle) is lengthened by not more than 10%. The spectral phase of the output pulses is smooth (dominated by dispersion) and can be well approximated with a polynomial function (dispersion coefficients: GDD $= 28\,\text{fs}^2$, TOD $= 37\,\text{fs}^3$, FOD $= -71\,\text{fs}^4$, specified at $\lambda_c = 1000\,\text{nm}$). Hence, good compression with standard tools is expected. A test with chirped mirrors designed for 800–1300 nm experimentally confirmed the compressibility of the pulses in at least this spectral range.

4.3.3   Discussion and Conclusions

The simplicity of the setup and the high conversion efficiency of up to 15% (with more than 80% of the generated pulse energy in the spectral range 700–1400 nm) make the cascaded DFG scheme an interesting PFS seed generation candidate for replacing the current cascaded HCF + XPW scheme. Moreover, the straightforward scalability suggests that when using the full output of the HCF in an adapted setup, seed energies of few tens of microjoule should be feasibly.

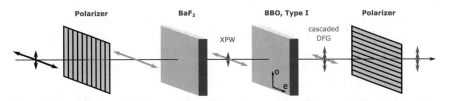

Fig. 4.16 Modified cascaded DFG scheme featuring an additional BaF$_2$ crystal for XPW generation and seeding of the cascaded processes in the subsequent BBO crystal

A potential issue of the setup arises from the fact that the seed for the cascaded DFG is linearly derived from the input pulses by rotation of the birefringent BBO crystal. As a consequence, any phase distortions or temporal artifacts in the input pulses are forwarded with this seed to the OPCPA chain, which—as has been already discussed—can negatively affect the amplified spectrum and spectral phase.

A simple modification to address this issue is the integration of an XPW generation stage as shown in Fig. 4.16. By installing a thin BaF$_2$ crystal in front of the BBO crystal, a cross-polarized wave is generated that serves as the ordinary polarized seed for cascaded DFG. Consequently, the roll-angle ψ of the BBO can be set to zero and thus the second polarizer that is adjusted along the ordinary axis of the BBO is now crossed with respect to the first polarizer. By this measure, temporal artifacts that do not take part in the nonlinear interactions cannot pass the setup anymore. Conveniently, the efficiency of XPW generation is of no particular importance in this scheme as it has been shown that less than 1% of the total energy on the o-polarized axis is sufficient for a decent cascaded DFG efficiency.

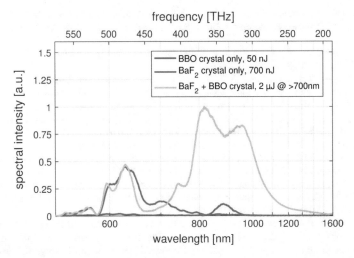

Fig. 4.17 Output spectra from cascaded DFG at a BBO orientation of $\psi = 0°$ where the ordinary polarized seed pulses are generated by XPW generation in BaF$_2$. The total input pulse energy is 60 μJ

In a short proof-of-principle experiment, a 500 μm BaF_2 crystal was installed few centimeters in front of the BBO crystal. Figure 4.17 depicts the measured spectra for 60 μJ input pulses. Since the additional dispersion from the BaF_2 crystal was not compensated, the cascaded DFG efficiency was not optimal. However, the obtained results demonstrate the feasibility of the concept.

Having a look at the cascaded DFG scheme from a conceptual point of view, it combines in a sense the two previously described alternative seed generation schemes and overcomes their shortcomings: the main drawback of the idler generation scheme is the angular chirp of the idler as a consequence of the non-collinear geometry required by phase matching. Working with a broadband pump and collinear beams, this problem is completely avoided in the cascaded DFG scheme. From the cascaded HCF + XPW generation scheme the new scheme inherits the HCF broadening and the XPW method for contrast enhancement. However, by employing only one fiber and a more efficient technique of wavelength conversion into the range of interest (700–1400 nm) an output energy increase by up to an order of magnitude seems feasible.

In summary, the combination of spectral broadening in a single HCF with XPW generation and cascaded DFG constitutes an attractive seed generation scheme that promises high-energy, high-contrast and ultra-broadband pulses with a smooth spectrum and spectral phase. Hence, the integration of an adapted version of this scheme into the PFS system is planned for the future.

References

1. A. Shirakawa, I. Sakane, T. Kobayashi, Pulse-front-matched optical parametric amplification for sub-10-fs pulse generation tunable in the visible and near infrared. Opt. Lett. **23**, 1292–4 (1998)
2. T. Kobayashi, A. Shirakawa, Tunable visible and near-infrared pulse generator in a 5fs regime. Appl. Phys. B **70**, S239–S246 (2000)
3. T.J. Wang, Z. Major, I. Ahmad, S. Trushin, F. Krausz, S. Karsch, Ultrabroadband near-infrared pulse generation by noncollinear OPA with angular dispersion compensation. Appl. Phys. B Lasers Opt. **100**, 207–214 (2010). https://doi.org/10.1007/s00340-009-3800-9
4. G. Pretzler, A. Kasper, K. Witte, Angular chirp and tilted light pulses in CPA lasers. Appl. Phys. B Lasers Opt. **70**, 1–9 (2000). https://doi.org/10.1007/s003400050001
5. H. Liebetrau, M. Hornung, A. Seidel, M. Hellwing, A. Kessler, S. Keppler, F. Schorcht, J. Hein, M.C. Kaluza, Ultra-high contrast frontend for high peak power fs-lasers at 1030 nm. Opt. Express **22**, 24776-24786 (2014). https://doi.org/10.1364/OE.22.024776
6. H. Fattahi, H.Wang, A. Alismail, G. Arisholm, V. Pervak, A.M. Azzeer, F. Krausz, Near-PHz-bandwidth, phase-stable continua generated from a Yb:YAG thin-disk amplifier. Opt. Express **24**, 24337 (2016). https://doi.org/10.1364/OE.24.024337
7. A. Jullien, J.-P. Rousseau, B. Mercier, L. Antonucci, O. Albert, G. ChEriaux, S. Kourtev, N. Minkovski, S. Saltiel, Highly efficient nonlinear filter for femtosecond pulse contrast enhancement and pulse shortening. Opt. Lett. **33**, 2353–2355 (2008). https://doi.org/10.1364/OL.33.002353
8. A. Buck, K. Schmid, R. Tautz, J. Mikhailova, X. Gu, C.M.S. Sears, D. Herrmann, F. Krausz, Pulse cleaning of few-cycle OPCPA pulses by cross-polarized wave generation, in *Frontiers in Optics 2010/Laser Science XXVI* (2010), pp. 8–9. https://doi.org/10.1364/FIO.2010.FMN2

9. L.P. Ramirez, D. Papadopoulos, M. Hanna, A. Pellegrina, F. Friebel, P. Georges, F. Druon, Compact, simple, and robust cross polarized wave generation source of few-cycle, high-contrast pulses for seeding petawattclass laser systems. J. Opt. Soc. Am. B **30**, 2607 (2013). https://doi.org/10.1364/JOSAB.30.002607

10. L. Canova, O. Albert, N. Forget, B. Mercier, S. Kourtev, N. Minkovski, S.M. Saltiel, R. Lopez Martens, R. Lopez Martens, Influence of spectral phase on cross-polarized wave generation with short femtosecond pulses. Appl. Phys. B **93**, 443–453 (2008). https://doi.org/10.1007/s00340-008-3185-1

11. G.P. Agrawal (ed.), *Nonlinear Fiber Optics* (Academic Press, 2001). https://doi.org/10.1002/adma.19900020919

12. S. Kourtev, N. Minkovski, L. Canova, A. Jullien, Improved nonlinear cross-polarized wave generation in cubic crystals by optimization of the crystal orientation. J. Opt. Soc. Am. B **26**, 1269–1275 (2009). https://doi.org/10.1103/PhysRevLett.84.3582

13. C. Skrobol, *High-Intensity, Picosecond-Pumped, Few-CycleOPCPA* (Ludwig-Maximilians-Universität München, PhDthesis, 2014)

14. H. Fattahi, A. Schwarz, S. Keiber, N. Karpowicz, Efficient, octave-spanning difference-frequency generation using few-cycle pulses in simple collinear geometry. Opt. Lett. **38**, 4216–4219 (2013). https://doi.org/10.1364/OL.38.004216

15. A. Kessel, S. A. Trushin, N. Karpowicz, C. Skrobol, S. Klingebiel, C. Wandt, S. Karsch, Generation of multi-octave spanning high-energy pulses by cascaded nonlinear processes in BBO. Opt. Express **24**, 5628 (2016). https://doi.org/10.1364/OE.24.005628

16. H. Tan, G.P. Banfi, A. Tomaselli, Optical frequency mixing through cascaded second-order processes in β-barium borate. Appl. Phys. Lett. **63**, 2472 (1993). https://doi.org/10.1063/1.110453

17. G.I. Petrov, O. Albert, N. Minkovski, J. Etchepare, S.M. Saltiel, Experimental and theoretical investigation of generation of a cross-polarized wave by cascading of two different second-order processes. J. Opt. Soc. Am. B **19**, 268 (2002). https://doi.org/10.1364/JOSAB.19.000268

18. J. Matyschok, T. Lang, T. Binhammer, Temporal and spatial effects inside a compact and CEP stabilized, few-cycle OPCPA system at high repetition rates. Opt. Express **21**, 475–479 (2013). https://doi.org/10.1063/1.123820.A

19. A. BaltuSka, T. Fuji, T. Kobayashi, A. Baltuska, T. Fuji, T. Kobayashi, Controlling the carrier-envelope phase of ultrashort light pulses with optical parametric amplifiers. Phys. Rev. Lett. **88**, 133901 (2002). https://doi.org/10.1103/PhysRevLett.88.133901

Chapter 5
OPCPA Experiments with Two OPA Stages

5.1 Performance of Alternative Seed Generation Schemes

In the following we report on the first series of OPCPA measurements which we conducted on two LBO stages in vacuum (cf. Fig. 3.9). The presented data have been taken together with Christoph Skrobol and have partially already been described in his thesis [1]. For the measurement campaign, the Tube served as the last pump amplifier and the pump compressor was still located in air. Due to the low damage threshold of a dichroic mirror inside the Tube, the output energy of the fundamental pulses was kept at 400 mJ. Unresolved thermal issues limited the repetition rate to 1 Hz. After compression and SHG, the energy of the frequency-doubled pump pulses was 4.2–4.5 mJ for the first and 75–81 mJ for the second OPCPA stage.

The seed pulses were provided by the new seed generation schemes described in the previous chapter, namely the idler generation and the cascaded HCFs + XPW generation.[1] While the idler scheme was only used for a short proof-of-principle experiment where OPA phase matching conditions have not been fully optimized, the data obtained with the XPW scheme represent the final outcome of a thorough optimization.

The OPA stages themselves consist of two LBO crystals (Type I, $\vartheta=90°$, $\varphi=14.5°$). Experimentally, an optimum crystal thickness of 4 mm in both stages was determined for highest OPCPA efficiency. While in principle two orientations of the crystals are possible (walk-off compensation or tangential phase matching [1, 2]), in practice the choice of orientation is only relevant for small beam sizes and thick crystals[2]: At a non-collinear angle of $\alpha_{int} = 1.1°$ and a walk-off angle of $\rho = 0.45°$ in both OPCPA stages, the relative lateral shift between pump and signal beams inside the LBO

---

[1] For this OPCPA campaign an early version of the XPW scheme was used that delivered only $\sim 1\,\mu J$ compared to $\sim 4\,\mu J$ in the final version that was described in Sect. 4.2.

[2] For some crystals and spectral bands also the suppression of parasitic SHG can play a role. For LBO and signal wavelengths from 700 to 1400 nm, however, none of the two orientations has an advantage over the other [3].

© Springer International Publishing AG, part of Springer Nature 2018
A. Kessel, *Generation and Parametric Amplification of Few-Cycle
Light Pulses at Relativistic Intensities*, Springer Theses,
https://doi.org/10.1007/978-3-319-92843-2_5

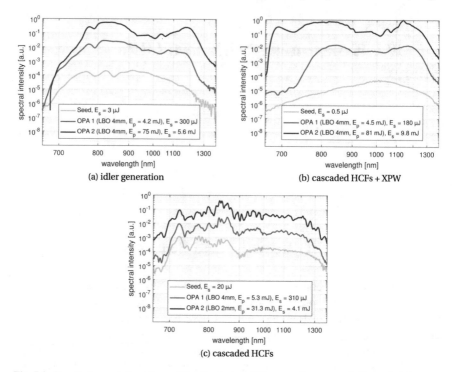

Fig. 5.1 Results from parametric amplification in two OPA stages that are seeded by **a** the idler generation scheme, **b** the cascaded HCFs plus XPW stage (both pumped by the Tube) or **c** the cascaded HCFs *without* XPW (pumped by the Booster). Figure (**c**) is identical to Fig. 3.8 in Sect. 3.3.1

crystals is $46\,\mu m$ (walk-off comp.) or $108\,\mu m$ (tangential phase matching). These values are small compared to the beam diameters of several millimeters and consequently we could not detect any difference in spectrum, energy or beam profile when testing both orientations. Since in theory, however, the group-velocity mismatch between pump and signal is slightly smaller in the tangential-phase-matching geometry [4], we installed the crystals in this orientation.

The amplification results shown in Fig. 5.1a, b demonstrate for both alternative seed generation schemes a broad amplification bandwidth with pulse energies of up to $\sim 10\,mJ$ (for XPW). For convenience, Fig. 5.1c shows once again the measurement results reported in Sect. 3.3.1 with the OPA stages in air, pumped by the Booster and seeded by the initial cascaded HCFs scheme.

Owing to the different conditions in the three measurements, the results are not entirely quantitatively comparable, however a few fundamental observations can be made:

- As intended, the implementation of both alternative schemes yields significantly *smoother spectra* for the input as well as for the amplified pulses compared to the initial seed generation scheme. This fact already indicates a reduced (if not

Table 5.1 OPCPA efficiencies measured with different seed generation schemes

Seed scheme	Efficiency 1st OPA stage	Efficiency 2nd OPA stage
Casc. HCFs (%)	5.5	12.2
Idler generation (%)	7.1	7.1
Casc. HCFs + XPW (%)	4.0	11.9

entirely suppressed) distortion of the spectral phase and therefore suggests a good temporal compressibility of the pulses.

- A comparison (see Table 5.1) of the measured efficiencies at the first OPA stage when seeded by the idler generation or the XPW scheme reveals that *higher seed energies lead to a better energy extraction from the pump* which agrees with intuitive expectations. In this respect, schemes with higher seed energy would be preferable. However, because of saturation and possibly back-conversion, this effect does not necessarily lead to higher output energies after the second OPA stage.

- It is interesting to note that the *residual angular chirp* of pulses generated with the idler scheme has no obvious adverse effect on amplification bandwidth in the first OPCPA stage[3] as can be seen in Fig. 5.1a. At first glance this observation might be surprising but can be understood considering two points: First, by imaging the idler beam from the position of generation onto the first OPCPA stage, any formation of a spatial chirp was prevented that could have affected the bandwidth at the center of the beam. And second, even though the non-collinear angle α is not identical for all wavelengths due to the angular chirp ($\alpha = \alpha(\lambda)$), the maximal deviation of $\pm 0.02°$ at the first and $\pm 0.005°$ at second stage over the full spectral range is too small to have a measurable effect on phase matching (cf. Fig. 2.14d). However, despite the good OPCPA performance, the problem of insufficient compressibility and focusibility of the amplified pulses because of the angular chirp remains. Therefore the idler scheme was eventually abandoned.

- Overall, the *OPCPA efficiencies* of 4.0–7.1% in the first and 7.1–12.2% in the second stage were significantly lower than predicted by simulations that suggest efficiencies of around 20% [1]. In fact, we achieved slightly better efficiencies in the high energy campaign presented in the next section but still not close to the theoretical predictions. On the other hand, a comparison with similar broadband OPCPA systems shows that efficiencies of 10–15% in high gain stages are rather the rule than the exception [5–8]. Hence, the simulations might neglect higher-order losses present in real world systems that prevent reaching higher efficiencies.

After these amplification tests, the XPW seed scheme was used to examine the compressibility of the amplified pulses. For this purpose, the pulses were coupled out of the OPCPA vacuum system after amplification. A fraction of the beam was forked off by a wedge, demagnified and sent through a 1″ chirped-mirror-compressor

---

[3]The amplification results from the second stage are less instructive as phase matching was not optimized in this measurement in contrast to the more extensive XPW-campaign.

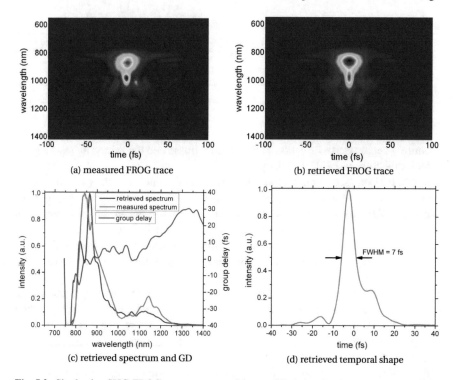

Fig. 5.2 Single-shot SHG-FROG measurements of the amplified signal pulses. **a** Measured FROG trace, **b** retrieved FROG trace, **c** measured and retrieved spectrum as well as retrieved group delay, **d** retrieved temporal shape. Pictures taken from [1] with the permission of the author

(16 reflections) and a pair of fused silica wedges. An additional neutral density filter further attenuated the beam to protect the diagnostics. After optimization of dispersion with the wedges and selecting a combination of three types of chirped mirrors (PC503, PC515, PC518: UltraFast Innovations), a compressed pulse duration of 7 fs (Fourier limit 6 fs) was measured with an early version of the SS-SHG-FROG as shown in Fig. 5.2.

The described measurements demonstrated for the first time the successful operation of a picosecond-pumped OPCPA system with multi-mJ output energy and a compressed pulse duration of slightly more than two optical cycles. However, in order to use the generated pulses for applications, the amplified pulse energy had to be further increased and at the same time the setup had to be modified to allow the temporal compression of the full beam (and not just an attenuated fraction). The efforts made to achieve these goals will be described in the following section.

5.2 High-Energy OPCPA Experiments

After the proof-of-principle experiments presented in the previous section, we will describe in the following the results of the OPCPA campaign at full pump energy available from the last pump amplifier.

5.2.1 Pump

Historically, the high-energy OPCPA campaign was started with the Tube as the last pump amplifier. Upon completion of the OPA test campaign described in the previous section its 400 mJ-limitation of output energy was overcome and 1 J maximal energy at 2 Hz repetition rate was provided. However, due to reoccurring damages of optics inside the amplifier and a corresponding degradation of the beam profile, the Tube was later replaced by the Cube that provided pulse energies of up to 1.2 J at 10 Hz on a daily basis (cf. Sect. 3.2). While most results presented in this section were obtained with the Cube, a few measurements still were taken with the Tube as will be indicated in the respective figures.

Compared to the status documented in [1], the rebuild of the pump compressor as described in Sect. 3.2 allowed to use the full energy of ∼1 J in the fundamental beam for compression and frequency-doubling to generate high-energy pump pulses for the OPCPA chain.

Figure 5.3 shows the beam profiles of fundamental and second harmonic of the Cube at the position of the frequency-doubling DKDP crystal at full energy. Even

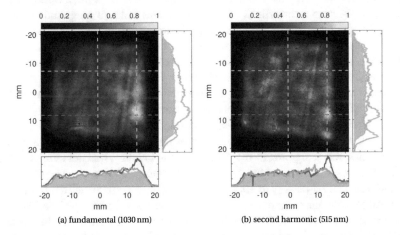

(a) fundamental (1030 nm)   (b) second harmonic (515 nm)

Fig. 5.3 Cube beam profiles of **a** the fundamental (1030 nm) and **b** the second harmonic (515 nm) at the position of the SHG crystal at full energy. Cross sections at different points are shown. Gray contours correspond to median values along the respective dimension

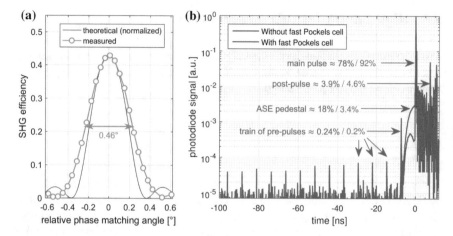

Fig. 5.4 **a** Acceptance angle of the 4 mm Type II DKDP crystal for SHG. **b** Time structure of the compressed fundamental pump pulses with and without the fast Pockels cell. Percentage values indicate the energy proportions relative to the total energy (Measurements taken with the Tube)

though some hot-spots are visible at the bottom right, the intensity distribution over the entire beam is quite homogeneous as shown by the cross sections with a total standard deviation of about 20% from mean.

The frequency-doubling efficiency was initially rather poor at about 45%, a fact already observed in [1] and suspected there to have its origin in wavefront distortions in the fundamental beam. By scanning the phase matching angle of the employed SHG crystal, however, we measured an acceptance angle of 0.46° on a 50% level (see Fig. 5.4a). This matches quite well the theoretical prediction by a simple analytical model [9]. Wavefront distortions with angles of this order of magnitude would be clearly visible in short distance beam deformation which we did not observe. This line of argument is also supported by the fact that the two minima in Fig. 5.4a at ±0.5° both reach a value of zero within the accuracy of measurement: in the presence of severe wavefront distortions, there would be a broad range of incident angles of the beam onto the crystal and therefore a perfect mismatch could not be achieved: the efficiency minima would show a clear offset from zero. So in summary, wavefront distortions could be excluded as the main source of low SHG efficiency.

Instead, we found that the time structure of the pump pulses was not as clean as expected and that a significant fraction of the total energy was not confined to the main pulse: measurements of the compressed 1 J pump pulses with a fast photodiode and different neutral-density filters for a high dynamic range revealed a series of small pre-pulses next to the main pulse, a strong post-pulse and an ASE pedestal as shown by the blue curve in Fig. 5.4b. These low-intensity features have a vanishingly small SHG efficiency and therefore reduce the overall efficiency. The features originate from different components of the pump chain.

The *pre-pulses* are created as leakage of the regenerative amplifier. For each round-trip, the finite contrast of out-coupling Pockels cell and thin-film-polarizer

Fig. 5.5 Record measurement of second-harmonic generation with compressed pulses from the Cube in a 4 mm DKDP crystal (Type II)

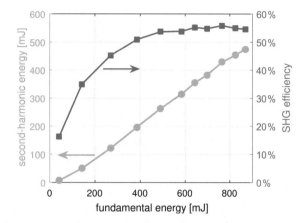

(TFP) leads to a small loss in pulse energy, forming a train of weak pulses that exit the cavity. However, as this pulse train carries less than 0.25% of the total energy, the effect can be ignored.

The *post-pulse* with about 4% of the total energy is created in a similar way: due to the limited contrast of Pockels cell and TFP, a fraction of the main pulse is not coupled out, makes one more amplified round-trip and exits the cavity almost undisturbed because of the comparably long fall-time of the Pockels cell. By replacing it with a version featuring a shorter fall-time, this artifact could be suppressed in the future.

Finally, the *ASE* contains approximately 18% of the energy and is already generated in the fiber-amplifier. In order to efficiently remove this pedestal and prevent its further parasitic amplification, the pulses have to be temporally cleaned before the stretcher where they are still short. For this purpose, a fast-switching Pockels-cell (Leysop Ltd, $\tau_{gate} \approx 200$ ps) was installed that successfully reduces the ASE energy to 3.4% and boosts the energy ratio in the main pulse from 78 to 92% (see Fig. 5.4b).

As a consequence, the SHG efficiency could be improved to 55%, corresponding to an increase in second-harmonic energy of 20%. In a record measurement, 860 mJ in the fundamental beam at the SHG crystal (1.2 J at the Cube and 75% transmission of compressor and beam line) were converted to 475 mJ in the frequeny-doubled beam (see Fig. 5.5). Typical energies at the two OPCPA stages 14 and 400 mJ respectively with a shot-to-shot energy stability of 1.5%. Figure 5.6 shows the fundamental and second-harmonic spectra at different amplifier output energies. The fact that the measured spectra in Fig. 5.6b are slightly narrower than theoretically expected might be an indication that either the temporal compression is not yet ideal or the used 4 mm DKDP crystal is too thick for the bandwidth of the pulses.

While the pulse duration of the fundamental beam was determined with an SHG-FROG [1], the duration of the second-harmonic pulses could be only indirectly measured: for this purpose, a narrow spectral band ($\Delta\lambda \sim 20$ nm) around 800 nm central wavelength was selected from the seed pulses to create a short probe pulse which was then temporally scanned relative to the pump pulse at the OPA stage.

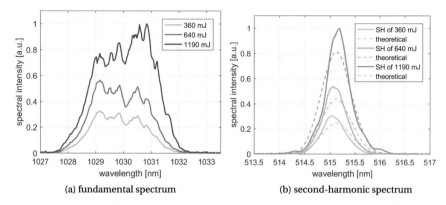

(a) fundamental spectrum   (b) second-harmonic spectrum

Fig. 5.6 **a** Output spectrum of the Cube at different energies. **b** Second-harmonic spectrum after frequency-doubling of the compressed pulses in 4 mm DKDP. The theoretical spectra (dashed curves) are calculated from the fundamental spectra in the left plot under the assumption of perfect compression and phase matching. Experimental spectra are scaled according to the contained energy, theoretical spectra such that the area under the curves is identical to the respective experimental curve

By recording the parametrically amplified output spectrum of the probe pulse for each delay, one obtains a correlation trace as shown in the inset of Fig. 5.7. In a separate measurement at time zero the scaling of the amplified signal energy with pump energy was found to be approximately linear for high energies. Therefore, the spectrally integrated correlation signal (green curve in Fig. 5.7) with a FWHM of 670 fs can be interpreted as the temporal intensity of the frequency-doubled pulses and provides an estimate for their duration.

Considering on the other hand the pronounced saturation of SHG that is visible in Fig. 5.5 at high energies, one would rather expect a comparable pulse duration of fundamental and second-harmonic pulses and thus a pulse duration of about 850 fs.

Depending on which value is used, the average pump intensities at the OPA stages are approximately 90 GW/cm^2 (70 GW/cm^2) at the first and 150 GW/cm^2 (120 GW/cm^2) at the second stage.

5.2.2   Signal

The seed/signal pulses for the high-energy OPCPA measurements were again generated by the cascaded HCFs + XPW scheme. However, in order to increase the seed energy compared to the proof-of-principle OPCPA campaign, an improved version of the XPW scheme was set up, delivering up to 4 μJ as presented in Sect. 4.2. As this improvement required a relocation of the XPW setup, the optical path length to the OPCPA stages and hence the effective dispersion was changed. Furthermore, the new OPCPA compressor was set up entirely in vacuum necessitating a new dispersion

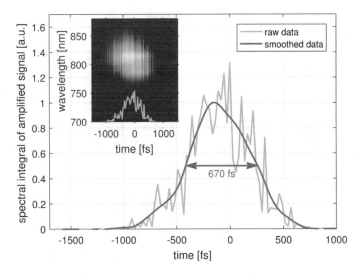

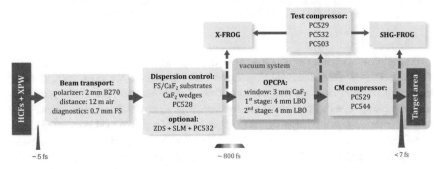

Fig. 5.7 Indirect measurement of the duration of the pump pulses (515 nm) by parametric amplifi-cation of narrowband, short signal pulses. The inset depicts the amplified signal spectra for different delays between signal and pump. The green curve in the inset as well as in the main plot shows the spectral integral for each delay. The width of this curve gives an estimate for the duration of the frequency-doubled pump pulses

Fig. 5.8 Scheme of the dispersion management scheme for the signal beam. FS = fused silica, PCxxx = chirped mirrors (dispersion curves shown in Fig. 5.9), ZDS = zero-dispersion stretcher, SLM = liquid-crystal spatial light modulator

management as shown in Fig. 5.8. In short, we introduce positive dispersion in front of the OPA stages to stretch the seed pulses to about 800 fs and compresses the ampli-fied pulses afterwards with negatively chirped mirrors. The compression results as well as details about the optional zero-dispersion stretcher (ZDS) and liquid-crystal spatial light modulator (LC-SLM) setup will be described in Sect. 5.2.3.

The motivation for adding chirped mirrors to the previously all-material based stretching scheme before the vacuum system was that in the OPCPA experiments conducted with the all-material scheme, a low OPA gain was measured for signal

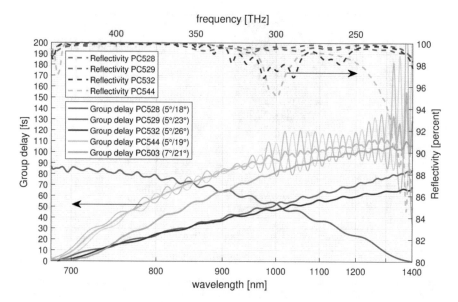

Fig. 5.9 Design curves of selected chirped mirrors used in this work. Except for PC503 the CMs are double-angle mirrors, i.e. all mirrors of one type are identical but have to be used in pairs with two different angles of incidence in order to collectively compensate the typical GD oscillations from the CM layer structure as shown exemplary for PC544. Group delay and reflectivity curves are design data "per reflection", i.e. the average of two mirrors placed at the given angles

wavelengths longer than ∼950 nm. This finding raised the question whether the pronounced TOD from material dispersion was responsible for the effect: as long-wavelength components are less stretched than short ones, there is a stronger competition for pump energy in this spectral region which in turn is expected to reduce the gain. In order to test the impact of this effect, a set of four chirped mirrors (PC528, UltraFast Innovations) was designed and installed to compensate the dispersion-induced TOD and generate a linear input chirp. Note, that the target GD-curve of the PC528 mirrors is designed to be monotonous as shown in Fig. 5.9 to avoid the error-prone production of TOD-only mirrors we suffered from previously [1]. An XFROG measurement of the stretched pulses revealed a perfect linear chirp as shown in Fig. 5.10.

In the end, the hypothesis that the amplified signal spectrum significantly depends on the TOD of the pulses proved experimentally false by testing two experimental scenarios: in the first case the seed was stretched by four reflections on PC528 mirrors as explained. In the second case the mirrors were bypassed and the seed was instead dispersed by transmission through a 9.5 mm thick fused silica substrate. By this means, the seed pulses feature the same group delay of ∼800 fs between spectral components at 700 and 1400 nm but with an increased TOD. Since both scenarios resulted in identical amplified spectra and energies within the accuracy of measurement, the constraint of a linear chirp was later given up to obtain an easily

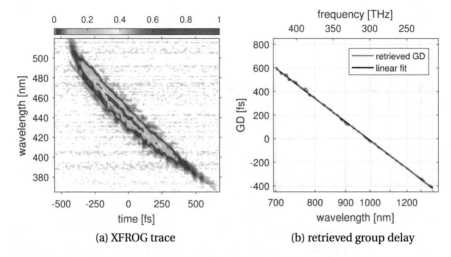

Fig. 5.10 **a** XFROG trace of the stretched signal pulses (optical path including 4x PC528 and 5 mm CaF$_2$). **b** Group delay retrieved from this trace featuring a linear chirp

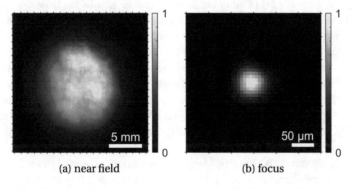

Fig. 5.11 **a** Signal beam profile at the position of the second OPCPA crystal. **b** Beam profile in the focus of a concave mirror with 1 m focal length

accessible additional degree of freedom for temporal compression by changing the number of PC528 reflections.

The near field profile of the signal beam at the position of the second OPCPA stage has a FWHM size of $10.5 \times 12\,mm^2$ and is shown in Fig. 5.11a. Focusing this beam with a concave mirror of 1 m focal length yields a spot size of $58 \times 55\,\mu m^2$ FWHM as shown in Fig. 5.11b which demonstrates the good focusibility (diffraction-limited size: $52 \times 49\,\mu m^2$).

5.2.3 OPCPA

5.2.3.1 Amplification Results

In Fig. 5.12 the results from parametric amplification are shown, where pump pulse energies of 13 and 415 mJ were used at the two OPA stages. In Fig. 5.12a, the typical spectra of the amplified signal pulses are plotted on a logarithmic scale, spanning the full range from 690 to 1350 nm with pulse energies of 1.0 mJ ($\sim$7% efficiency) after the first and 53 mJ ($\sim$13% efficiency) after the second stage. Despite the more pronounced spectral modulations compared to the low energy OPA campaign, the overall smoothness of the spectra is still acceptable given the large bandwidth.

One exception is the spike around 1140 nm which is better seen when plotting the normalized spectra on a linear scale (cf. Fig. 5.12b). The origin of this spike was initially unclear as it could not be detected in the unamplified seed. In a later conducted investigation it was found to be created by a weak train of post-pulses in the seed at this specific wavelength, each about four orders of magnitude lower than the main pulse and separated by $\sim$200 fs. As the stretched main seed pulse temporally overlaps with just the central part of the pump pulse, one or two of these post-pulses have exclusive overlap with the trailing wing of the pump and hence experience large gain when driving the OPCPA stages in saturation. By adjusting phase matching and/or timing between signal and pump, the spike can be reduced, however, at the expense of amplified bandwidth and energy. Due to the short separation between the pulses, we suspect that the post-pulses originate from a dielectric coating in the beam path of the seed (an erroneous reflection from windows or the backside of mirror substrates can be excluded as 200 fs separation would correspond to a substrate thickness of only $\sim$20 μm). The investigations to identify and replace the problematic component are still ongoing.

Figure 5.12c shows that the amplified spectrum after the second stage theoretically supports a Fourier-limited pulse duration of 4.2 fs, corresponding to 1.6 optical cycles of the electric field at 817 nm central wavelength. In this ideal case, 77% of the total energy is confined to the central peak. Excluding the presumably incompressible spectral components at 1100–1160 nm (i.e. the discussed spike), the limit increases to 4.6 fs.

In Fig. 5.12d, the amplified beam profile at full energy is displayed where the imprint of pump features onto the originally smooth seed profile (cf. Fig. 5.11a) is clearly visible. The elliptical shape is caused by a mismatch of the pulse fronts of pump and signal which originates from two independent sources: first, owing to the non-collinear geometry, pump and signal beam intersect at an angle α_{int} resulting in a pulse front mismatch in the horizontal plane. And second, due to a slight mis-alignment in the pump compressor,[4] the pulse front of the pump pulses themselves

---

[4]In the current setup, this misalignment is difficult to correct as it emerges only when the compressor chamber is evacuated, i.e. when the components are inaccessible for manual alignment. In a currently ongoing rebuild this issue is solved by a motorization of the critical components. For details about the influence of compressor misalignment on pulse front tilt, see also Sect. 6.3.

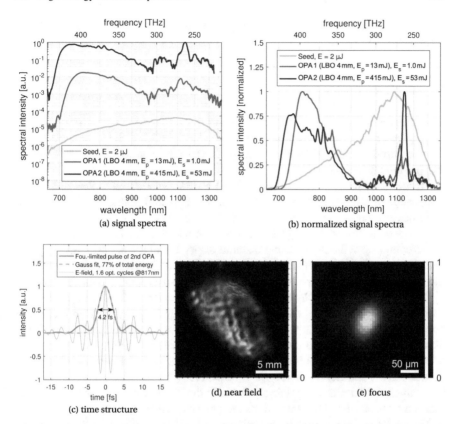

Fig. 5.12 Spectra of seed and amplified signal. **a** shows the energy-scaled spectra on a logarithmic plot, **b** shows the same spectra on a linear scale and normalized to unity for direct comparison **c** is the theoretical time structure for perfectly compressed pulses, **d** is the near field beam profile after the second OPCPA stage and **e** shows the beam profile in the focus of a concave mirror with 1 m focal length

are tilted in the vertical dimension with respect to the direction of propagation. The superposition of these two effects yields a diagonal mismatch angle between pump and signal pulses. Since both pulses are short and their lateral extension is large, this mismatch results in a poor spatio-temporal overlap at the edges of the beam profiles and leads to the elliptical shape. A detailed quantitative analysis of the pulse front mismatch as well as a scheme for compensation can be found in Sect. 6.3.

The poor overlap also results in a reduced total amplification efficiency at the second OPA stage: in a later series of measurements presented in Sect. 6.2.2 an efficiency of up to 18% was observed when limiting the analysis to the central part of the beam. Therefore, we expect a significant increase of total amplified signal pulse energy when the pulse fronts are properly matched.

As can be seen in Fig. 5.12e, the ellipticity of the beam profile additionally affects the focus spot size: at $81 \times 59 \, \mu m^2$ FWHM along the major and minor axes, the focus area is increased by about 50% with respect to the focus area of the unamplified beam (cf. Fig. 5.11b).

5.2.3.2 Compression

The compression of chirped, few-cycle pulses down to their Fourier-limit is a challenging task by itself. Furthermore, as described in Sect. 2.2.1, the parametric amplification introduces an additional spectral phase (the OPA phase) that depends on gain and phase-matching conditions.

Hence, after the first proof-of-principle OPCPA experiments it was planned to implement an acousto-optic modulator (Dazzler, FastLite) in the signal path before amplification for adaptive dispersion control. The first design of an appropriate scheme was based on a single Dazzler and an additional grism-compressor [10, 11] to compensate the unavoidable material dispersion of the acousto-optic crystal. Due to the spectral range and the broad bandwidth of the PFS signal pulses, however, the expected total throughput of such a scheme was unacceptably poor at ~1%.

The idea of a second approach developed in cooperation with FastLite was to split (and afterwards recombine) the signal pulses in a Mach-Zehnder-interferometer-like setup into two color channels with two individual Dazzler units. By this means, the Dazzlers would be able to compensate their own material dispersion and thus no additional compressor would be needed. Due to this fact and because of a better adaption of the Dazzlers to their respective wavelength bands, the expected throughput of this solution was acceptable at ~10%. In the end, however, the increased costs induced by the second Dazzler unit resulted in a cancellation of the project.

Hence, it was necessary to temporally compress the amplified signal pulses again by chirped mirrors only. Since after the cancellation of the Dual-Dazzler-purchase there was no other possibility to shape higher-order dispersion terms, we decided to design two different sets of chirped mirrors: one set to compensate a perfectly linear chirp (PC529, cf. Fig. 5.9) and one set to compensate dispersion of the OPA crystals and the predicted OPA phase (PC531). Thus, by exchanging pairs of one type of mirrors with the other and by fine-tuning the residual GDD with CaF_2-wedges, one gains a certain flexibility for adjusting higher-order dispersion.

At first, a measurement campaign with $1''$ mirrors in air (dispersion-compensated by one more set of mirrors: PC532) was performed to examine the design and production of the planned full-size $2''$ mirrors and to verify the compressibility of the unamplified signal pulses. The campaign unveiled the typical challenges of all-chirped-mirror compression of few-cycle pulses:

- A production error in the PC531 mirrors was detected that introduced strong phase-modulations which only became evident as an accumulated effect and was not noticed e.g. when single mirrors were analyzed in a white-light interferometer. Therefore, the PC531 mirrors in the test compression setup had to be replaced by an older set of mirrors (PC503) with a similar dispersion.

• It proved very difficult to determine the spectral target phase for the chirped mirrors with the required precision since systematic errors of few percent in the XFROG measurements of stretched pulses or in the dispersion calculations with Sellmeier equations can already result in a critical deviation from the real spectral phase of the pulses and eventually prevent full compression. Moreover, because of the necessarily large number of chirped mirrors the same applies for the precision of production: small systematic discrepancies in the multi-layer structure can lead to a significant accumulated deviation of the introduced spectral phase from the target curves. These deviations can be of higher order and thus a compensation with e.g. CaF$_2$ wedges is in general not possible. Hence, the initially measured chirp of the signal pulses as well as the accordingly calculated target curves of the chirped mirrors were used just as a starting point for the compressor setup before iteratively testing different mirror combinations to achieve final compression.

In the end, a combination of $6 \times$ PC529, $14 \times$ PC532 and $4 \times$ PC503 was experimentally found to yield optimal compression of the unamplified pulses. Figure 5.13a, b show the corresponding XFROG and SS-SHG-FROG traces with retrieved pulse durations of 5.7 fs and 6.0 fs respectively. When parametric amplification was switched on, the pulses were not compressed anymore by this mirror combination as expected due to the additional OPA-phase described in Sect. 2.2.1. This effect is clearly visible from the FROG measurements shown in Fig. 5.13c, d. In order to determine the OPA phase as precisely as possible, the difference in the group delay of amplified and unamplified pulses was retrieved from the FROG traces. Additionally, a spectral interferometry measurement was conducted. For this purpose, a fraction of the seed pulses was split off, guided parallel to the OPA system and interfered afterwards with the unamplified and amplified signal pulses. The resulting spectral modulations and retrieved group delays are shown in Fig. 5.13e.

The experimental results were compared to simple theoretical predictions based on the quasi-constant ratio between OPA phase and phase mismatch for high gain stages[5] as already suggested by Fig. 2.7b. As can be seen in Fig. 5.13f there is reasonable agreement between experiment and theory. Hence, the OPA phase determined by SI was included in the design of the target curve for a new chirped mirror (PC544).

The PC544 as well as the already tested PC529 design were produced on 2″ substrates to support the full beam size of the amplified pulses and formed the final compressor. A reduction of beam size prior to compression as an alternative option was excluded for damage threshold reasons: testing a demagnification by a factor of two of the quasi-compressed signal pulses to emulate a 1″ chirped mirror compressor resulted in mirror surface damages already at ∼20 mJ pulse energy.

Figure 5.14 shows the compact compressor assembly and beam path inside an old vacuum chamber which we adapted for our purposes. The chamber itself is located on a separate optical table (cf. Fig. 3.9) and connects the OPA system with the beam line to the target area of our lab.

---

[5]For the phase mismatch, the results from Fig. 2.14b were used. The gain was assumed to be 10^3 in the first and 50 in the second stage. The crystal thickness was 4 mm in both stages.

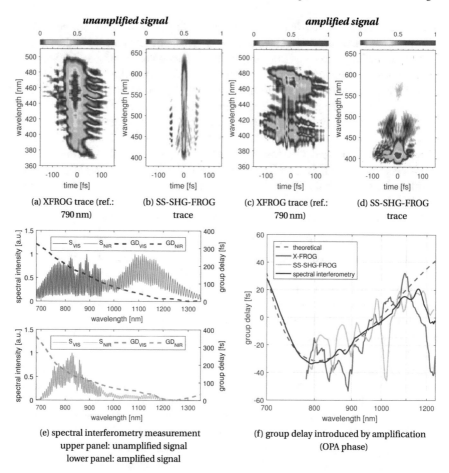

Fig. 5.13 Compression of unamplified signal pulses in air and measurement of OPA phase. Installed chirped mirrors: 6 × PC529, 14 × PC532 and 4 × PC503. **a, b** FROG traces of unamplified pulses. **c, d** Corresponding traces for amplified pulses (after two OPA stages). To protect diagnostics and chirped mirrors, amplified pulses were attenuated with wedges. **e** Measured interference spectra of reference pulses bypassing the OPA system with seed and amplified pulses passing through the system. Dashed curves show the retrieved group delays. **f** comparison of simulated and measured group delays introduced by amplification (OPA phase)

Despite testing several combinations of chirped mirrors[6] including changing the number of "positive" PC528 mirrors in front of the OPCPA stages, the compression of amplified pulses was not satisfactory and measured pulse durations were not shorter than ~10 fs, almost a factor of two longer than the Fourier limit. Hence, we decided to borrow a ***liquid-crystal spatial light modulator*** (***LC-SLM***, SLM-S640,

---

[6] ... a cumbersome process involving venting and re-evacuation of the vacuum chambers for each modification as well as the partial rebuild of the compressor setup due to different design angles of the chirped mirrors.

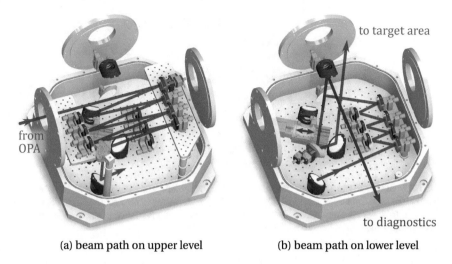

(a) beam path on upper level   (b) beam path on lower level

Fig. 5.14 Vacuum chamber with 2″ double-angle chirped mirrors for compression of the amplified signal pulses. A motorized stage is used to pick the compressed beam for measurement of pulse duration (purple beam path)

JENOPTIK) from a neighboring group to gain a finer control over dispersion [12]. Implementing such a device was first ruled out because of concerns that the imprint of the inherent pixelation on the spectrum and phase could deteriorate the contrast. A careful theoretical investigation, however, yields that temporal artifacts due to this pixelation are not stronger than 10^{-6} relative to the main pulse. This could be experimentally confirmed by a measurement of the contrast of the unamplified pulses at least down to 2×10^{-5} (for details see Appendix A.5). As the LC-SLM is installed before OPA and as the temporal artifacts expected from simulation lie outside the OPA pump window of ∼1 ps they are not further amplified. This reduces the possible impact of the LC-SLM on the final contrast to the acceptable level of 10^{-10} assuming a parametric gain of about 10^5.

To use an LC-SLM for spectral shaping, it has to be located in a plane where the wavelengths of the pulse to be shaped are spatially separated. To this end, a so-called "zero-dispersion stretcher" was build that creates this separation. The setup is schematically shown in Fig. 5.15 and consists of a blazed grating and a spherical mirror. The scheme earns its name from the fact that it does not introduce different optical path lengths for different spectral components as normal stretchers do. Thus, except for the unavoidable propagation through the LC-SLM (∼6 mm FS) and air, there is no additional ("zero") dispersion. The setup was integrated in the signal beam path before the OPCPA system with four additional reflections on PC532 mirrors (not shown) to roughly compensate the material and air dispersion. The mirror configuration in the compressor chamber after OPA was $6 \times$ PC529 and $8 \times$ PC544.

In the optimization process, the spectral phase of the amplified pulses was changed with the LC-SLM while observing the temporal compression live with the SS-SHG-

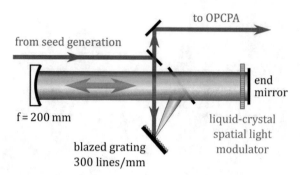

Fig. 5.15 Sketch of the zero-dispersion stretcher setup consisting of a blazed grating to angularly disperse the signal pulses, a spherical mirror to collimate the wavelengths and an end mirror to send the beam back for recombination. A liquid-crystal spatial light modulator in the Fourier plane is used to adjust the spectral phase of the seed pulses for OPCPA

FROG. Finally, this yielded a pulse duration of 6.4 fs as shown in Fig. 5.16 and corresponds to a waveform featuring just 2.2 optical cycles at the measured central wavelength of 870 nm. One might notice that in Fig. 5.16c the measured and retrieved spectra do not exactly match. As a consequence, the retrieved pulse duration as well as the Fourier limit of 5.9 fs might slightly underestimate the real pulse duration, however, by not more than approximately 10%.

At a total chirped mirror compressor transmission of 84% (about 7% of the 16% losses are from silver mirrors for beam transport), there remains a compressed pulse energy of up to 45 mJ. Taking into account that the main peak of the temporal intensity curve contains 75% of the total fluence (cf. Fig. 5.16d), the generated pulses feature a peak power of 4.9 TW.

Guiding the amplified pulses to the target chamber further losses of $\sim$17% due to beam transport have to be considered. Focusing the beam with a 45° off-axis parabolic mirror (f/1.2) resulted in a focal spot size of $2.0 \times 3.2 \,\mu m^2$ FWHM as shown in Fig. 5.17. Taking into account the spatial intensity distribution, the peak intensity on target can be as high as 4.5×10^{19} W/cm^2. In terms of the normalized vector potential, this corresponds to a relativistic value of $a_0 = 5.0$.

5.2.3.3 Contrast

In the introduction, the potentially high temporal contrast was listed as one of the key advantages of OPA systems over conventional laser amplifiers when it comes to the generation of relativistic light pulses. To test the validity of this expectation for our system, we measured the contrast of the signal pulses amplified in both OPA stages with a third-order autocorrelator (Tundra, UltraFast Innovations). The device splits the input beam into two arms, generates second-harmonic pulses in one of them and then recombines these pulses with the fundamental in the other arm. By recording

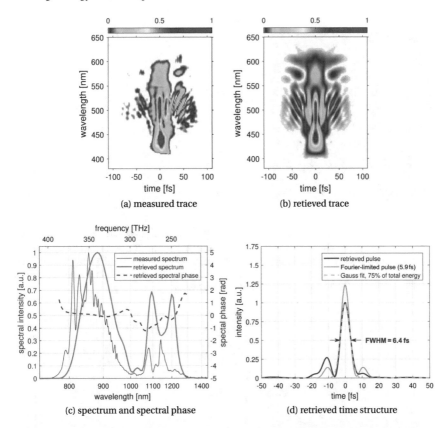

Fig. 5.16 Compression results of amplified signal pulses with 2″ CMs in vacuum. Dispersion of seed pulses was optimized with the LC-SLM. Temporal characterization was performed with the SS-SHG-FROG located in air adding 0.7 mm fused silica and 500 mm air to the dispersion. **a, b** Measured and retrieved FROG traces (G-error = 2%, 256 × 256 pixels). **c** Measured and retrieved spectrum and spectral phase. **d** Time structure of the retrieved pulse and the Fourier-limit pulse

the produced third-harmonic pulses with a photo-multiplier-tube while scanning the delay between the two arms, one obtains the temporal structure of the input pulses.

Initially, the Tundra was used with the full OPA bandwidth of 700–1300 nm. However, in this configuration the measured background level of the device was raised by about two orders of magnitude compared to the ideal case, most likely due to higher-order nonlinear effects in the crystals. Since anyway the provided Tundra did not support the full OPA bandwidth by design due to phase-matching limitations of the nonlinear crystals and bandwidth limitations of the mirrors, a bandpass filter was installed at the entrance of the device to cut out the accepted spectral range ($\Delta\lambda = 40$ nm at $\lambda_c = 800$ nm). By this means the full dynamic range of the device could be exploited.

Figure 5.18 shows that with this optimized configuration of the setup we could measure an excellent ratio of about 10^{11} between the main pulse and the background.

(a)   **(b)**

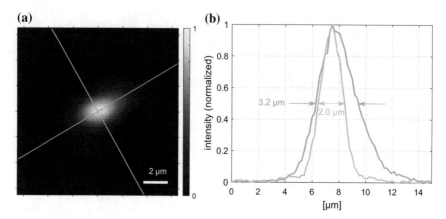

Fig. 5.17 **a** Focus spot of the amplified and temporally compressed signal beam on target, generated with a 45° off-axis parabolic mirror (f/1.2). For the measurement the beam was attenuated and imaged with an infinity-corrected microscope objective (40x, NA 0.65) and a lens (f = 200 mm) onto a CCD. **b** shows the intensity profile along the major and minor axes as indicated in (**a**). Raw data by courtesy of Olga Lysov and Vyacheslav Leshchenko

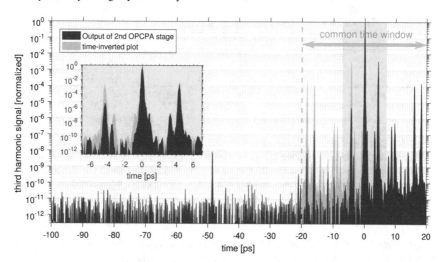

Fig. 5.18 Contrast measurement with a third-harmonic autocorrelator (Tundra, UltraFast Innovations). The blue curve shows the third harmonic of the signal pulses after OPA in two stages. The gray curve is a copy of the blue one, mirrored at time zero to allow the identification of pre-puls artifacts generated by post-pulses. For details see the main text

In particular, it can be seen from the inset that there is no long pedestal around time zero and that the intensity stays at the noise level until ~1 ps before the main peak (temporal resolution of the Tundra: ~100 fs).

In the plot, a few pre-pulses are visible which we believe are measurement arti-facts: in general, the idea of a third-order autocorrelator is to exploit the nonlinearity of the SHG process in order to obtain "clean" frequency-doubled pulses that can be used in the following to scan the fundamental pulses. However, a temporal feature in the input pulses with an intensity ratio of $1/Q$ with respect to the main peak is not entirely removed by SHG but just attenuated, yielding a ratio of $1/Q^2$ in the second harmonic beam. During the THG scan, these attenuated features in the second-harmonic arm temporally overlap with the main peak of the fundamental and create time-inverted artifacts in the recorded curve. Hence, post-pulses in the input beam lead to the measurement of pseudo pre-pulses. Comparing the third-harmonic signal in Fig. 5.18 with the time-inverted plot (gray curve) within the common window of ± 20 ps, the correspondence of pre- and post-pulses is evident. Since furthermore the ratio between coinciding features agrees in most cases quite well with the expected $1/Q^2$-scaling, we conclude that the measured pre-pulses are not real or at least far less intense than they appear to be. In a later measurement we could identify plane-parallel substrates which we use for dispersion control as responsible for some of the post-pulses. The vacuum exit window on the diagnostic beam path to the Tundra as well as the OPA-crystals are other potential sources.

Due to the bandwidth limitation of the device to the region $\sim$780–820 nm, in principle we cannot exclude the existence of temporal features from spectral components outside this range. However, as all spectral components of the signal pulses are synchronously generated and follow the same optical path in the system, we believe that the scenario of wavelength-selective pre-pulses or coherent pedestals is rather unlikely.

5.2.4 Summary

In summary, in this chapter the successful parametric amplification of octave spanning pulses in two OPA stages was presented, yielding pulse energies of more than 50 mJ. By temporal compression of the stretched amplified pulses with a combination of two types of chirped mirrors and an LC-SLM for fine dispersion control, a pulse duration of 6.4 fs was achieved corresponding to a 2.2-cycle waveform at 870 nm central wavelength. At a compressed pulse energy of $\sim$45 mJ, the achieved peak power was up to 4.9 TW. A focus measurement with an off-axis parabolic mirror showed that the peak intensity on target can be as high as 4.5×10^{19} W/cm^2. These values mark a notable improvement compared to the prior low-energy OPA campaign described briefly at the beginning of the chapter: While in this campaign an energy of 400 mJ (at 1030 nm) from the last pump amplifier was used to provide signal pulses with a compressed peak power of 0.65 TW, tripling the fundamental pump energy to 1.2 J in the high-energy campaign resulted in a signal peak power more than seven times higher than before. Hence, not only the absolute energies were scaled but additionally the overall efficiency of the PFS system could be more than doubled. The measured excellent temporal contrast of $\sim 10^{11}$, finally, confirmed the usability of the generated pulses for applications.

Currently, first high-harmonic-generation experiments on solid surfaces are carried out that use the described OPA system. In parallel, preparations were started for an upgrade of the PFS, which aims at upscaling the amplified pulse energies by more than order of magnitude in a third OPA stage. These preparations will be described in the next chapter.

References

1. C. Skrobol, *High-Intensity, Picosecond-Pumped, Few-CycleOPCPA* (Ludwig-Maximilians-Universität München, PhDthesis, 2014)
2. F. TrAger (ed.), *SpringerHandbook of Lasers andOptics* (Springer, Berlin, Heidelberg, 2012) https://doi.org/10.1007/978-3-642-19409-2
3. J. Bromage, J. Rothhardt, S. Hadrich, C. Dorrer, C. Jocher, S. Demmler, J. Limpert, A. Tunnermann, and J.D. Zuegel, Analysis and suppression of parasitic processes in noncollinear optical parametric amplifiers. Opt. Express **19**, 16797–16808 (2011). https://doi.org/10.1364/OE.19.016797
4. C. Homann, Optical parametric processes to the extreme: fromnew insights in first principles to tunability over more than 4 octaves, Ph.D. thesis, Ludwig-Maximilians-Universität München, 2012
5. M. Baudisch, B. Wolter, M. Pullen, M. Hemmer, J. Biegert, High power multi-color OPCPA source with simultaneous femtosecond deep-UV to mid-IR outputs. Opt. Lett. **41**, 3583 (2016). https://doi.org/10.1364/OL.41.003583
6. A. Harth, M. Schultze, T. Lang, T. Binhammer, S. Rausch, U. Morgner, Two-color pumped OPCPA system emitting spectra spanning 1.5 octaves from VIS to NIR. Opt. Express **20**, 3076–81 (2012)
7. S. Fang, G. Cirmi, S. H. Chia, O.D. Mucke, F.X. Kartner, C. Manzoni, P. Farinello, G. Cerullo, Multi-mJ parametric synthesizer generating two-octave-wide optical waveforms, in *Pacific Rim Conference on Lasers and Electro-Optics, CLEO-Technical Digest* (2013), pp. 4–5.https://doi.org/10.1109/CLEOPR.2013.6600174
8. J. Matyschok, T. Lang, T. Binhammer, Temporal and spatial effects inside a compact and CEP stabilized,few-cycle OPCPA system at high repetition rates. Opt. Express **21**, 475–479 (2013). https://doi.org/10.1063/1.123820.A
9. F.-T. Wuand, W.-Z. Zhang, Consideration of angular acceptance angle in BBO crystal on a highly efficient second harmonic generation. Opt. Laser Technol. **30**, 189–192 (1998). https://doi.org/10.1016/S0030-3992(98)00032-2
10. T.H. Dou, R. Tautz, X. Gu, G. Marcus, T. Feurer, F. Krausz, L. Veisz, Dispersion control with reflection grisms of an ultra-broadband spectrum approaching a full octave. Opt. Express **18**, 27900–27909 (2010). https://doi.org/10.1364/OE.18.027900
11. A. Kastner, Dispersion control of ultra-broadband light pulses, Master thesis, Ludwig-Maximilians-Universität München, 2013
12. A.M. Weiner, Femtosecond pulse shaping using spatial light modulators. Rev. Sci. Instrum. **71** 1929–1960 (2000). https://doi.org/10.1063/1.1150614

Chapter 6
Preparations for a Third OPA Stage

At the present state as described in the last chapter, the PFS is able to deliver 4.9 TW, few-cycle pulses which are already used for applications. To boost the peak power further towards the envisaged petawatt level enabling experiments with unprecedented pulse parameters, an upgrade of the PFS system is currently under construction. The core element of this upgrade is a third OPA stage whose operation requires major modifications to the current system. In the following, the most important considerations and experimental tests that were carried out to prepare for this step will be described.

6.1 General Layout

In Fig. 6.1, the planned layout of the upgraded PFS system is shown. In total we aim for an increase of the amplified signal pulse energy by more than an order of magnitude which requires an increase of pump pulse energy by about the same factor.

To achieve this, an upscaled version of the Cube amplifier is currently under development. It reuses the established 20-pass relay imaging design [1] with an enlarged beam size allowing for higher energies from the pump diodes. By installing two (instead of one) separately pumped Yb:YAG crystals in both image planes of the setup, the heat aggregation in the crystals is reduced and cooling is more efficient. From simulations and experimental tests with the Cube, the new amplifier (dubbed here "Cube10") is expected to deliver pulse energies of more than 10 J at 10 Hz.

In the current design it is planned to split the generated pump pulses before compression into a 1 J-channel for the first two OPCPA stages and a 9 J-channel for the third stage. On the one hand, this design has the drawback that both channels have to be individually compressed requiring a tight packing of mirrors on two layers in the compressor chamber to allow the simultaneous use of the large diffraction grating.

© Springer International Publishing AG, part of Springer Nature 2018
A. Kessel, *Generation and Parametric Amplification of Few-Cycle*
Light Pulses at Relativistic Intensities, Springer Theses,
https://doi.org/10.1007/978-3-319-92843-2_6

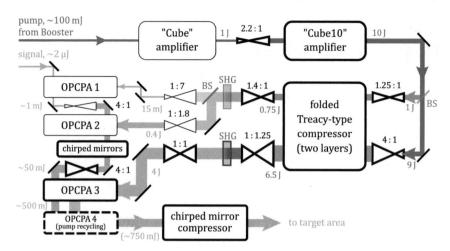

Fig. 6.1 Layout of the extended PFS OPCPA system. Elements drawn with a thick frame correspond to new or modified components. The magnifications of existing or planned telescopes as well as the estimated pulse energies at certain points are specified

On the other hand, the design has a few advantages: First, splitting the pulses only after compression would imply a transmission of high-intensity pulses through the (thick) beam splitter substrate introducing B-integral. Second, in case of an outage or a longer service of the 10 J-amplifier, with the planned scheme there remains the possibility to bypass the 10 J-amplifier and beam splitter and to continue experiments with the Cube on the 1 J-channel (i.e. operating two OPCPA stages). And finally, the separate beam paths in the compressor allow to individually control compression and the pulse front tilt of the pump pulses for all three OPCPA stages as will be explained in Sect. 6.3.

In general, it is planned to attach the third OPCPA stage including pump-SHG to the existing system, i.e. to leave the setup of first and second OPA stage (as presented in Fig. 3.9) largely unchanged. One modification, however, becomes necessary: as there is only limited space on the diffraction grating of the compressor and as fluences should be kept as low as possible to protect the grating from damage, the beam size of the 1 J beam has to shrink compared to the current setup to leave enough space for the 9 J beam. The telescopes before the compressor will therefore scale the 1 J and 9 J beams to fill the maximum free apertures in the compressor of about 30 mm and 100 mm respectively. The telescope that images the compressed 1 J pulses (∼0.75 J transmitted) onto the SHG crystal will correspondingly be adapted to yield the same beam size and fluence as in the current system. The scaling of the 9 J beam after compression is determined by the aperture of the nonlinear crystals and will be discussed in the next section.

The signal pulses will be partially re-compressed between the second and third OPA stage by four chirped mirrors to counteract the dispersion of the first two OPA crystals. A magnifying telescope adapts the signal beam size for the third stage which will be pumped by approximately 4 J at 515 nm. Based on our experience with the current OPA stages (compare also the tests in Sect. 6.2.2), we expect a total efficiency of 12–15% for the third OPA stage corresponding to amplified signal pulse energies of 500–600 mJ. After compression by chirped mirrors (4″ diameter), this yields a peak power of 60–70 TW on target, assuming a pulse duration of 6.5 fs and a 80% transmission through the compressor and the beamline.

This peak power could be even further increased by reusing the remaining pump energy after the third OPA stage to amplify the signal once more in a fourth stage ("pump recycling"). To allow for this measure, the vacuum chambers for the extended system have already been designed to accommodate another OPA crystal on a second layer. Simulations and experimental studies of pump recycling by other groups suggest an increase of the output energy by 50–70% by this technique [2–4]. Therefore, an ultimate peak power of 90–120 TW might be achievable with the upgraded PFS system.

6.2 Evaluation of Nonlinear Crystals

In the original concept of the PFS, it was planned to use DKDP crystals for frequency-doubling of the pump as well as for the OPA stages owing to the commercial availability of this crystal type in large sizes [5]. While other crystals such as LBO are known to be superior to DKDP,[1] they could not be considered for the high-energy stages of the PFS in the past due to their limited apertures (max. 50 mm diameter LBO available on an experimental basis in 2007). Only at the first two OPA stages of PFS where beam diameters are comparably small and crystals easily available, DKDP could be replaced by LBO as has been described in [6].

Today, crystal production has evolved and LBO crystals can be purchased with diameters of up to 80 mm (Cristal Laser S.A.) at slightly higher cost than DKDP crystals of the same size. Hence, both crystal types are possible options for SHG and OPA in the planned upgrade.

To calculate the fluences to be expected when working with crystals of this size, we assume a circular 6th-order Gaussian pump beam profile for the output of the 10 J-amplifier and a clear aperture of 75 mm for the crystals. Requiring a geometrical transmission of 95% of the total energy through this aperture yields a FWHM beam diameter of 65 mm (cf. Fig. 2.2a). At this beam size, the peak fluence is ~0.19 J/cm^2 for the fundamental pump pulses if we assume an energy of 6.5 J after compression. Aiming for an SHG efficiency of 60%, this corresponds to a peak fluence of ~ 0.12 J/cm^2 at 515 nm.

To test whether an operation of SHG and OPA stages at these fluences is possible, the damage threshold of AR-coated samples of both DKDP (Eksma) and LBO (Cristal Laser) was measured.

---

[1] The higher nonlinearity of LBO compared to DKDP enables at an identical gain the use of thinner crystals and therefore yields a lower B-integral and better phase-matching conditions which result in a broader amplified bandwidth.

Table 6.1 Results from damage threshold measurement. Listed values indicate the highest fluence the crystals could withstand for at least a few hundred shots. Increasing the fluence further (by $\sim$20%), a damage could be detected on the front surface

	1030 nm	515 nm
DKDP, EKSMA 65 × 4 mm, $\vartheta = 54.4°$, $\varphi = 0°$	2.3(5) J/cm^2	0.5(1) J/cm^2
LBO, Cristal Laser 20 × 25 × 3.2 mm, ϑ=90°, $\varphi = 0°$	4.3(9) J/cm^2	1.0(2) J/cm^2

6.2.1 Damage Threshold Measurements

For the measurement of damage threshold, the compressed fundamental or frequency-doubled pulses from the Tube were cropped by an iris and focused with a lens ($f = 1$ m) onto the crystals, yielding an airy disk with 180 μm FWHM.[2] The pulse energy was then increased step by step while observing the crystal under scrutiny by camera and eye.

In general, when ramping up the energy we first detected damages at the back side of the crystals, then bulk damage at $\sim$1.5x higher energy and finally damages at the front side at $\sim$23x higher energy. Since the B-integral value in the test setup was up to 10 and since the beam diameter was small compared to the crystal thicknesses (cf. Table 6.1), we blame self-focusing to be responsible for this effect. This is in agreement with the observed emergence of strong diffraction patterns when imaging the transmitted beam at high energy onto a camera (still below damage threshold). Therefore we believe that the damage threshold of the front surface is the most meaningful quantity as the fluences inside the crystal or at the backside depend on the magnitude of self-focusing and could not be reliably determined.

At 2 Hz repetition rate of the Tube, several hundred shots were recorded on different spots on the samples. Table 6.1 shows the highest peak fluences the crystals could withstand before the emergence of scattering from front surface damage. The measured value for LBO at 515 nm could be compared with more precise test data from the manufacturer (1.3 J/cm^2 at 1 kHz, 1.5 ps), showing good agreement within the error of measurement and by considering the longer pulse duration in their setup.[3]

Comparing the measured damage thresholds of the test crystals with the planned peak fluence of 0.12 J/cm^2 at 515 nm, the safety margin (at least for DKDP) appears rather small: if we assume that inhomogeneities in the pump beam profile (cf. Fig. 5.3a) result in a locally higher fluence by a factor of two with respect to the nominal value and if we consider a further increase due to self-focusing by another factor of

---

[2]To achieve the same focal spot size for both wavelengths, the diameter of the iris was adjusted.

[3]As a test, we temporally stretched our pump pulses by about a factor of 4 resulting in an increase of the damage threshold fluence by a factor of $\sim$2. Interestingly this result is in accordance with the $\sqrt{\tau}$ scaling which is supposed to apply only for pulses longer then few tens of picoseconds [7].

two for the back surface,[4] the effective fluence can be as high as $\sim 0.5\,\text{J/cm}^2$. Thus, DKDP cannot be safely used in the described scenario. LBO on the other hand at a twice higher damage threshold would be still acceptable.

Apparently, the damage threshold deficiency of DKDP could be circumvented by purchasing larger crystals and working at lower fluences. However, enquiries to different manufacturers revealed that prices of DKDP crystals larger than 100 mm exceed those of 80 mm LBO crystals. This fact, together with the generally infe-rior nonlinear properties of DKDP therefore renders this option rather unattractive. Hence, we decided to employ LBO crystals for both the third (and fourth) OPA stage as well as for SHG of the pump pulses.

6.2.2 Determination of Optimal Thickness

SHG crystal

For the experimental determination of optimal crystal thickness, Cristal Laser pro-vided us several small LBO samples (Type I, $\varphi = 13.3°$). To test SHG with high fluence, the compressed pump pulses from the Cube were demagnified with a 3:1 reflective Galilean telescope. By cropping the fundamental as well as the frequency-doubled beam behind the crystal with a single common iris (3.8 mm diameter), a homogeneous region of the beam profile was selected to emulate a flat-top-like beam while still covering a representative area of the crystal. Fundamental and second-harmonic pulses were separated by three high-reflective mirrors coated for 515 nm which allowed to determine the respective pulse energies.

Figure 6.2 shows the measured SHG efficiencies in dependence on the input flu-ence for different crystal samples. The maximum achievable efficiency is similar for all crystals, however, as expected thinner crystals reach this point and saturate at higher input fluences. Furthermore, for the 2, 2.5 and 3 mm samples there exists a clear maximum beyond which back-conversion leads to a drop in efficiency. Due to damages on the back side of the first tested 3 mm crystal and the high-reflective mirrors that occurred when increasing the input fluences to more than $0.2\,\text{J/cm}^2$ the following crystals were tested only up to this point. The 1 mm crystal (being the last sample in the campaign) was then again measured at high fluences up to $0.26\,\text{J/cm}^2$ without any damages on the crystal. Therefore, we assume that the reason for damaging of the 3 mm sample was self-focusing inside the crystal which is more severe for thicker crystals. It should again be mentioned at this point that the pump beam was demagnified for the measurement and inhomogeneities that remained after spatial filtering were consequently smaller than in the full beam, leading to faster

---

[4]Note that the self-focusing-induced damage threshold ratio of 2–3 between front and back surface which we observed in our measurements represents an upper limit: as we spatially filter the pump beams in the real system, hot-spots smaller than ~ 1 mm are removed and self-focusing is less dramatic compared to the 180 μm beam size used for damage threshold measurements.

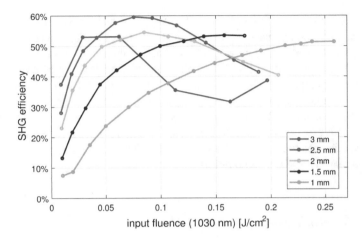

Fig. 6.2 Measured SHG efficiency for LBO crystals (Type I, $\varphi = 13.3°$) of different thickness

self-focusing. Hence, the effect should be less pronounced in the final setup and the crystals might survive higher fluences.

For the assumed 6th-order Gaussian pump beam from the last amplifier the peak fluence of the fundamental beam at $0.19\,\text{J/cm}^2$ corresponds to a weighted mean fluence[5] of $0.16\,\text{J/cm}^2$. At this value, the 1.5 mm crystal shows best performance. Even though Fig. 6.2 suggests a slightly lower peak SHG efficiency compared to e.g. the 2.5 mm sample, we believe that achieving an efficiency of close to 60% is still feasible with the 1.5 mm crystal: as the high-reflective mirrors used for measuring the frequency-doubled energy were not in the best shape and degraded further during the campaign, the measured absolute values of second harmonic energy and efficiency are not exact and presumably slightly lower than the actual values.

OPA crystal
Originally, it was intended to experimentally determine the optimal crystal thickness for the third OPA stage in the final setup with the new 10 J-pump amplifier. This would have had the advantage that no assumptions needed to be made regarding the exact output energy of this amplifier, transmission through the system or SHG efficiency, and crystals could be tested under real conditions. By growing the 80 mm crystals already beforehand, the time between final thickness specification with small test samples and delivery of the large crystals could be minimized.

---

[5]By "weighted mean fluence" we refer to the average fluence value of a beam, weighted with the generated second-harmonic energy: For any non-flat-top beam shape, there is a range of fluences F within the clear aperture of the crystal which can be described by a fluence distribution $\mathcal{D}(F)$, normalized to $\int_0^\infty \mathcal{D}(F)\,\mathrm{d}F = 1$. Assuming a constant SHG efficiency, it is $F_{\text{SHG}} \propto F$ and one obtains for the energy distribution (sloppily: the "energy fraction transported by the respective fluence"): $E_{\text{SHG}}(F) \propto \mathcal{D}(F)\,F$. Hence, the weighted mean fluence is $\bar{F} = \frac{\int_0^\infty E_{\text{SHG}}(F)\,F\,\mathrm{d}F}{\int_0^\infty E_{\text{SHG}}(F)\,\mathrm{d}F} = \frac{\int_0^\infty \mathcal{D}(F)\,F^2\,\mathrm{d}F}{\int_0^\infty \mathcal{D}(F)\,F\,\mathrm{d}F}$. For a 6th-order Gaussian beam and a crystal aperture of $1.15 \times$ FWHM this yields $\bar{F} = 0.82\,F_{\text{max}}$ ($0.88\,F_{\text{max}}$ for the non-saturated case where the SHG efficiency increases linearly with F).

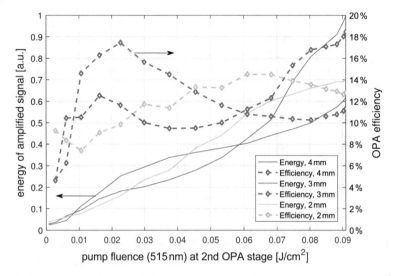

Fig. 6.3 Measured energies and efficiencies of the second OPCPA stage for three different LBO crystal thicknesses

However, since the time of completion of the currently developed 10 J-pump amplifier is hard to predict, it was decided to perform a few measurements with the present system consisting of two OPA stages. By this means, one obtains already a first estimate of the crystal performance to be expected in the final setup. As a boundary condition for the measurements, concurrent experiments with the system were not be disturbed for longer than few days. Therefore, the OPCPA scheme was kept unchanged and only the crystal of the second stage was replaced by different samples (a time consuming procedure by itself due to the vacuum-based setup). The downside of this compromise was that the used pump fluence and gain did not fully resemble those of the planned third stage.

To suppress any influence of the pulse front mismatch between pump and signal pulses (compare the distorted beam profile in Fig. 5.12d) both beams were cropped after the OPA stages with an iris and measured pulse energies and spectra at the middle of the beam. Three different LBO crystals with large apertures were used in the second OPA stage: two crystals with 40 mm diameter and a thickness of 4 and 3 mm (Crystal Laser) as well as a $25 \times 25 \, mm^2$ crystal with 2 mm thickness (EKSMA). Note, that we decided not to install the small test samples used in the previously described SHG campaign in order to avoid cropping of the pump beam inside vacuum which would result in material ablation and possible damages of optics.

The results shown in Fig. 6.3 were obtained keeping the output of the Cube as well as the amplification at the first OPA stage constant. The effective pump fluence for the second stage was controlled by rotating the polarization of the pump pulses with a half-wave plate. It is obvious from the plot that the scaling of the OPA efficiency with pump fluence strongly depends on the thickness of the installed crystal. As expected from theory and similar to the SHG results presented in the last section, the two thicker crystals show a local efficiency maximum at low pump fluences (around $0.02 \, \text{J/cm}^2$) before back-conversion leads to a drop in efficiency. For higher fluences, the efficiency of the 4 mm crystal increases again, surpasses the value of the first maximum and peaks at about 18% for the highest used pump fluence of $0.09 \, \text{J/cm}^2$. At this configuration, also the largest amplified signal energy was measured. Note that this maximization of energy and efficiency was the reason why the 4 mm crystal was used for OPA experiments in [6] and Chap. 5.

In contrast, the first efficiency maximum of the 2 mm crystal is located at $0.065 \, \text{J/cm}^2$ and hence at significantly higher fluence values than for the other two crystals. Energy-wise, at maximum pump fluence the crystal ranks second after the 4 mm sample. Slight back-conversion is visible at this point which is expected to be advantageous for the stability of the OPA stage as theoretically discussed in [6]. We could confirm this prediction experimentally as energy fluctuations of only 5% were measured for the 2 mm crystal which was about a factor of two smaller than for the 4 mm crystal (11%).

To gain a better understanding of the amplification behavior of the crystals, a series of measurements was performed where the pump fluence was ramped up and the amplified signal spectrum was recorded for each point. The obtained results are shown in Fig. 6.4. As one can see, the energy and efficiency scaling of Fig. 6.3 is generally reproduced, however, there are significant differences depending on the examined spectral region: In Fig. 6.4a–c the already discussed oscillatory behavior of the efficiency of the 4 mm crystal is visible at 850 and 900 nm. An extreme case is the region around 950 nm where almost full depletion can be observed at $0.05 \, \text{J/cm}^2$. For wavelengths longer than 950 nm, the development of efficiency with increasing pump fluence is heterogeneous: at 1000 nm it continuously drops, at 1100 nm and 1200 nm it remains approximately constant.

For the 2 mm LBO crystal, the situation is quite different: the efficiencies in all spectral regions scale very similarly (Fig. 6.4d–f) and reach their maxima at or slightly before the highest measured pump fluence. From the raw data points (thin lines in the plots) also the better stability compared to the 4 mm crystal is apparent.

The amplified signal spectra for the three crystals at $0.09 \, \text{J/cm}^2$ pump fluence are displayed in Fig. 6.5a. It is obvious that the 4 mm as well as the 3 mm crystal provide more energy in the long-wavelength part of the spectrum, while the 2 mm crystal results in an overall significantly smoother spectrum. This is reflected by the theoretical time structure of the perfectly compressed pulses as shown in Fig. 6.5b: 4 mm and 3 mm crystals allow shorter pulse durations whereas the 2 mm crystal leads to reduced side wings and a better confinement of energy in the main peak. In practice, we expect that the difference in pulse duration is less severe than the numbers suggest as the mentioned spectral modulations might locally affect the spectral phase and therefore prevent full compression to the Fourier limit.

Making a final decision on optimal crystal thickness for the third OPCPA stage based on the presented data is not straight forward. From the point of view of OPA efficiency and achievable peak intensity, the 4 mm crystal appears to provide best

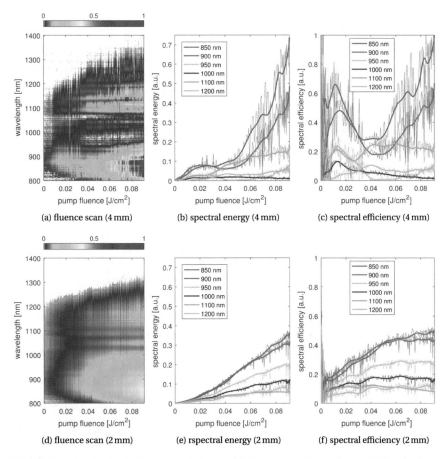

Fig. 6.4 Results of a pump fluence scan at the second OPA stage with 2 and 4 mm LBO crystals. **a**, **d** show the recorded spectra of the amplified signal pulses in dependence on the fluence as a color coded plot. **b**, **e** show the spectral energy for a few selected wavelengths. **c**, **f** show the corresponding efficiencies. Thin lines are raw data, thick lines are smoothed curved. The plots are scaled according to the pulse energies to allow a direct comparison of the 2 mm and the 4 mm crystal

results. However, since the pump fluence could not be increased up to the planned value of $0.12\,\text{J/cm}^2$ in the performed measurements, it is yet unclear how the OPA efficiency curve continues in this region. Therefore, the 3 mm crystal, recovering from its local efficiency minimum at $0.08\,\text{J/cm}^2$, might be the better choice to maximize the OPA efficiency at full pump energy.

On the other hand, working far beyond the first local efficiency maximum—as for the 4 and 3 mm crystals—has disadvantages in terms of spectral smoothness (and thus compressibility) and energy fluctuations as discussed. In this respect, the 2 mm crystal performs better than the thicker samples. The assumption that these factors play an important role is supported by first SHHG experiments performed in our

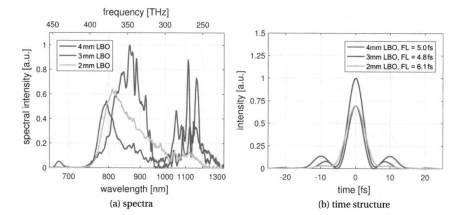

Fig. 6.5 **a** Measured spectra of the signal pulses after parametric amplification in three different LBO crystals at 0.09 J/cm² pump fluence. Spectra are scaled according to their respective pulse energy. **b** Corresponding theoretical time structure of the pulses when compressed to their Fourier-limit

group where best results were obtained with the 2 mm crystal. This observation, however, has to be considered as preliminary since the intensity on target was just above the threshold of SHHG. More precise data are expected from measurements that will be conducted after a rebuild of the pump compressor, beam line and target chamber. Promising higher pulse energies on target and a more reliable SHHG, these reconstructions will allow a thorough optimization of laser parameters including a test of different crystals.

Another aspect that needs to be considered is that the gain in the tested second OPA stage was about 20–30 (cf. Fig. 6.3) while it is expected to be only about 10 in the third stage. Due to the higher signal energy at the input, in low-gain OPA stages the energy is extracted faster from the pump and the point of saturation is reached after a shorter propagation distance in the crystal compared to stages with high gain [6]. This aspect once again calls for the use of thinner crystals (potentially even 1.5 mm) in the third OPA stage.

Summary

In summary, the *SHG* measurements with LBO crystals of different thickness yield a conclusive picture: At the target fluence of 0.16 J/cm² in the fundamental beam, the 1.5 mm crystal provides the highest conversion efficiency of all samples and therefore also the highest frequency doubled pulse energy. Hence this will be the thickness of choice for the 80 mm diameter SHG crystal to be ordered. In case that the currently developed last pump amplifier does not reach (or slightly exceeds) the desired output energy of 10 J, the imaging telescopes before and after the pump compressor (cf. Fig. 6.1) could be adapted to keep the fluence at the SHG crystal at the target value.

For the **OPA** crystal the situation is not so clear: first, the performed measurements with the present system did not fully resemble the conditions at the future third OPA stage. And second, there are different parameters that could be used as the figure of merit: In terms of amplified energy, OPA efficiency and spectral width, the 4 mm crystal performs best, whereas the 2 mm crystal is preferable in terms of stability, spectral smoothness and cleanness of the temporal shape.

More information—especially regarding the presumably most important figure of merit, the peak intensity—is excepted from a second series of measurements which is planned for the near future.

6.3 Pulse Front Matching

For phase matching reasons, the OPCPA stages of the PFS system employs a non-collinear geometry, i.e. signal and pump beams intersect at an angle α_{int} in the crystals. As a consequence of this non-collinearity and the fact that the intersecting pulses are short (pulse duration $\tau < 1$ ps), there is only a limited region of good spatio-temporal overlap in the center of the beams as depicted in Fig. 6.6a.

For constant τ and α_{int}, the effect of poor spatio-temporal overlap becomes more pronounced with growing beam diameters D_{OPA} resulting in a laterally cropped amplified signal beam. Moreover the combination of pulse front mismatch and temporal chirp of the signal pulses leads to a spatial chirp in the amplified beam. In the first OPCPA stage of the PFS system this effect is still small but at the second, larger stage it is already measurable. For the planned third stage, finally, the pulse front mismatch must be compensated to prevent an unacceptable beam deformation and low OPCPA efficiency. This compensation can be achieved by tilting the pulse front of the pump[6] by an angle β_{int} to reduce the mismatch angle $\gamma = \alpha_{int} - \beta_{int}$ as shown in Fig. 6.6b.

For a quantitative estimation, we start at an OPCPA simulation performed in [6] with the parameter set shown in Table 6.2. In Fig. 6.7, the distorted amplified beam profile as well as the spatial chirp (indicated by the simulated spectra at different points along the non-collinear dimension) are clearly visible. Based on this simulation, we define the maximum tolerable separation between pulse fronts (cf. Fig. 6.6) as

$$h_{max} := 15\,\mu m, \tag{6.1}$$

i.e. a factor of five smaller than in the simulated case. The corresponding tolerable mismatch angle is:

---

[6]In principle one could also tilt the signal pulse front but since this beam is to be used for further experiments and since a PFT implies angular dispersion and distorts the beam focus, this is not an option for the PFS system.

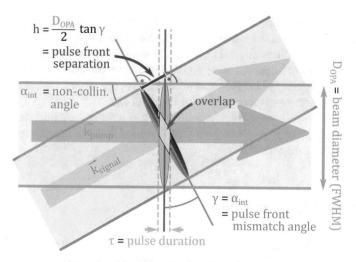

(a) pulse fronts perpendicular to k-vectors

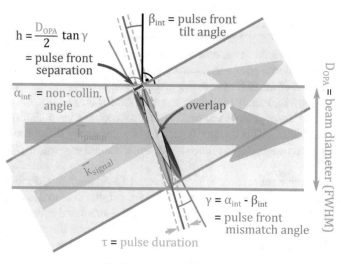

(b) tilted pump pulse front

Fig. 6.6 a Schematic illustration of poor spatio-temporal overlap when short pulses with large diameter cross at a non-collinear angle. **b** By tilting the pulse front of one beam the overlap can be improved

Table 6.2 Simulation parameters for non-collinear OPCPA with non-matched pulse fronts (derived from [6])

$$\tau = 1\,\text{ps}$$
$$D_{\text{OPA}} = 8\,\text{mm}$$
$$\gamma = \alpha_{\text{int}} = 1.1^\circ$$
$$h = \frac{D_{\text{OPA}}\,\tan\gamma}{2} \approx 77\,\mu\text{m}$$

Fig. 6.7 Simulated beam profile and spectra at different points of the beam (taken from [6] with the permission of the author)

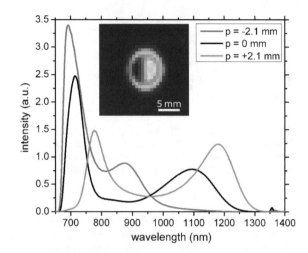

$$\gamma_{\text{max}} = \text{atan}\left(\frac{2\,h_{\text{max}}}{D_{\text{OPA}}}\right) = \text{atan}\left(\frac{30\,\mu\text{m}}{D_{\text{OPA}}}\right) \approx \frac{30\,\mu\text{m}}{D_{\text{OPA}}} \tag{6.2}$$

As this angle decreases for larger beam diameters D_{OPA}, at some point γ_{max} becomes smaller than the angle between signal and pump pulse fronts, i.e. the non-collinear angle α_{int}. From this point on, it is necessary to tilt the pulse front of the pump beam to reduce the mismatch angle such that $\gamma < \gamma_{\text{max}}$. The target PFT angle is given by the non-collinear angle

$$\beta_{\text{int}} \overset{!}{=} \alpha_{\text{int}} \tag{6.3}$$

with a tolerable deviation of

$$\Delta\beta_{\text{int}} = \gamma_{\text{max}}. \tag{6.4}$$

In the following, we will discuss how to generate such a PFT for all three OPCPA stages.

6.3.1   PFT Control with Transmission Gratings

In general, the pulse front of a beam can be tilted by introducing an angular chirp [8] which can be accomplished in many different ways. A simple and adaptive method

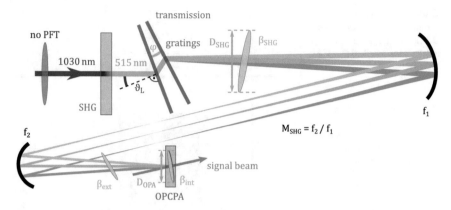

Fig. 6.8 Schematic setup of a pair of transmission gratings that introduce angular dispersion and therefore tilt the pulse front of the pump pulses after frequency-doubling for optimal overlap with the signal beam in the non-collinear OPCPA. For a better visibility the spatial chirp resulting from angular dispersion between the two gratings is omitted

is to send the beam through a pair of transmission gratings and control the PFT by adjusting the relative angle between these gratings [9]. This method was also our first choice to generate a PFT in the pump beam.

As the pump pulses in the PFS system are imaged from the SHG crystals onto the OPCPA stages (cf. Fig. 3.9) the most suitable position for the transmission gratings would be close to the SHG crystal as shown in Fig. 6.8 since in this case any spatial chirp created by propagation is compensated by the imaging system. The required pulse front tilt angle β_{SHG} at the SHG crystal position can be calculated as:

$$\beta_{ext} = \arcsin\left(n_{LBO,515\,nm}\ \sin\beta_{int}\right) \qquad\qquad \text{PFT angle at OPA crystal}$$
$$(6.5)$$

$$= \arcsin\left(n_{LBO,515\,nm}\ \sin\alpha_{int}\right) \approx n_{LBO,515\,nm}\ \alpha_{int} \qquad (6.6)$$

$$\beta_{SHG} = M_{SHG}\ \beta_{ext} \approx M_{SHG}\ n_{LBO,515\,nm}\ \alpha_{int} \qquad \text{PFT angle at SHG crystal}$$
$$(6.7)$$

where M_{SHG} is the magnification factor of the telescope between SHG crystal and OPCPA stage and $n_{LBO,515\,nm} \approx 1.6$ is the extra-ordinary refractive index of LBO at the second-harmonic pump wavelength. The tolerance

$$\Delta\beta_{SHG} = M_{SHG}\ n_{LBO,515\,nm}\ \Delta\beta_{int} = M_{SHG}\ n_{LBO,515\,nm}\ \frac{30\,\mu m}{D_{OPA}} \qquad (6.8)$$

defines the maximum permissible deviation from β_{SHG}.

For an optimum installation of the transmission gratings, two angles are important:

(a) The angle of incidence of the beam onto the gratings should match the Littrow angle ϑ_L (shown in Fig. 6.8) at which gratings typically show highest diffraction efficiency. It is defined as:

$$\vartheta_L = \arcsin\left(\frac{\lambda_0}{2d}\right) \tag{6.9}$$

where λ_0 is the central wavelength and d is the grating constant. If the grating pair is not parallel, an incident angle of ϑ_L cannot be accomplished for both gratings at the same time. In this case, the total diffraction efficiency is maximized if both angles are symmetrically offset from ϑ_L. For small tilt angles between the gratings, however, the effect is negligible. Hence, we will assume in the following that the first grating is hit under the Littrow angle and the second one is slightly off.

(b) The tilt angle φ between the gratings determines the PFT. For the generation of a small PFT angle β, it can be analytically approximated (compare [8]) by:

$$\varphi \approx \frac{\cos\vartheta_L}{\tan\vartheta_L}\frac{d}{\lambda_0}\beta = m_\varphi\,\beta \tag{6.10}$$

where we define $m_\varphi := \frac{\cos\vartheta_L}{\tan\vartheta_L}\frac{d}{\lambda_0}$ as a scaling factor.

The corresponding linear dependence between β and φ also holds for the tolerances:

$$\Delta\varphi = m_\varphi\,\Delta\beta_{SHG} \tag{6.11}$$

The value of $\Delta\varphi$ not only defines the required accuracy of grating alignment but also sets a limit to deformations of the grating substrates themselves. This effect is not negligible as the substrates have to be large to transmit the pump beam and thin to avoid B-Integral due to high intensities. Therefore a deformation—especially because of the one-sided AR-coating—is very likely.

Figure 6.9a shows as a simple example the case of a spherical bending of one of the gratings. As the angle φ between the gratings is not constant anymore but varies across the beam profile ($\varphi = \varphi(x)$), this leads effectively to a bent pulse front. If this bending is severe, it causes a poor spatio-temporal overlap of signal and pump pulses in the OPA crystal similar to the already discussed case of plane pulse fronts that intersect at an angle. To avoid this poor overlap, the local angle between the gratings must not deviate from φ by more than $\Delta\varphi$ across the whole beam profile.[7] For the sagitta s (cf. Fig. 6.9a) this implies

---

[7]This approach slightly overestimates the effective mismatch as most pump energy is confined to the central part of the beam where bending angles are smaller than in the outer parts.

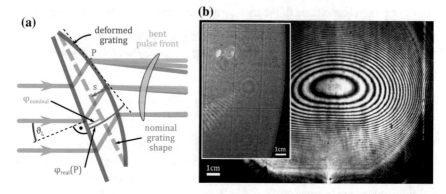

Fig. 6.9 **a** Sketch of a grating pair where one grating is tilted and spherically bent leading to a tilted and bent pulse front of the transmitted pulse. **b** Measured interferogram of a flat reference with the 0°-reflection of a thin test grating indicating an elliptical bending of the grating substrate. Inset: rectangular stitching artifacts from grating production

$$s \overset{!}{\leq} \frac{\Delta\varphi \, D_{\text{proj}}}{2} = \frac{\Delta\varphi \, D_{\text{SHG}}}{2 \cos \vartheta_L} = s_{\max} \tag{6.12}$$

where $D_{\text{proj}} = D_{\text{SHG}}/ \cos \vartheta_L$ is the projected beam size on the gratings.

In order to evaluate the severity of this effect for real gratings, measurements with a representative test sample from LightSmyth (150 mm diameter, 0.95 mm thickness, and 1702 lines/mm) were performed. To examine the *flatness of the substrate*, we used a large-aperture imaging Michelson interferometer ($\lambda = 780$ nm) at our institute and interfered the direct 0°-reflection from the wafer with the reflection from a flat reference.[8] The resulting interferogram in Fig. 6.9b shows an elliptical bending with a maximal height difference of ~30 μm across the entire wafer and about 13 μm in the area approximately covered by the beam ($D_{\text{proj}} \approx 70$ mm). Compared to the threshold of $s_{\max} \approx 47$ μm calculated with Eq. (6.12), this value is acceptable, even more as the main source of deformation is presumably the AR-coating: if both gratings feature a similar deformation, the PFT bending is partially compensated in the final setup when the gratings are oriented such that the ruled sides face each other (compare the considerations by Zhou et al. in [10]).

In general, thin transmission gratings with high diffraction efficiencies for 515 nm that match our size requirements are difficult and costly to manufacture. For this reason, LightSmyth uses projection photolithography to write comparably small 20×50 mm^2 grating patches on a wafer and arranges several of these patches to generate a quasi-continuous large grating area ("*stitching*") [11]. However, even

---

[8]It should be noted here that the common test setup with a double pass through the grating at Littrow angle does *not* provide meaningful information regarding surface flatness as in this configuration any small deformation is intrinsically compensated and does not appear in the interferogram (for details about his effect see [10]).

though the border between patches is less than $10\,\mu$m wide, the mesh-like structure is well visible. This can be seen in the inset of Fig. 6.9b where a beam that passed the grating twice was imaged onto a CCD camera. Moving the camera out of the image plane or trying to spatially filter the artifacts resulted in a strong enhancement of the effect, suggesting interference of neighboring patches as the source (probably due to a relative shift between the patches that does not exactly match the groove period). The accordingly expected distortions in the pump beam profile at the OPCPA crystal as well as in the amplified signal would not be desirable but probably still tolerable.

As a showstopper, however, we identified the low ***damage threshold*** of the gratings. In a series of measurements a small uncoated test sample from LightSmyth was exposed to focused, temporally compressed 515 nm pump pulses and revealed an upper fluence limit before damage of $0.056(20)\,$J/cm^2, more than a factor of two smaller than the planned fluence of about $0.12\,$J/cm^2. This low damage threshold can be explained to some extent by local field enhancements due to the grating structure of theoretically up to a factor of four in intensity (according to LightSmyth). However, since a reference measurement with an uncoated BK7 substrate survived more than $0.65\,$J/cm^2 (the highest possible intensity on the day of measurement) there might be also other reasons like micro-defects or the like.

A measurement of the damage threshold at 1030 nm with the same test grating yielded a significantly higher survival intensity of $0.34(6)\,$J/cm^2, though still critically close to the planned intensity of $0.19\,$J/cm^2 of the fundamental beam at the SHG crystal. Hence, an alternative scheme based on tilting the pulse front of the *fundamental* beam with transmission gratings *before* frequency-doubling proved to be not a viable option either.

In summary, the low damage threshold of the tested transmission gratings rendered them unusable for our purposes. Based on the assumption that gratings from other manufacturers would show similar performance, it was decided to look for an alternative solution.

6.3.2 PFT Control by Adjustment of the Pump Compressor

In many laser systems angular dispersion is accidentally generated when a beam passes a grating compressor that is misaligned in the sense that the grating surfaces or grooves are not parallel to each other. Usually the PFT resulting from this non-parallelism is unwanted, however, the effect can also be exploited to generate a desired PFT by controlled misalignment [12, 13]. Compared to the previous scheme with transmission gratings, this method has the advantage that no additional components have to be installed and hence losses are avoided.

In the folded design of the PFS pump compressor [14, 15] with just one grating, a horizontal and a vertical roof mirror (RM), the described non-parallelism can be realized by a misalignment of the horizontal RM. An interesting feature of the setup is that if the overall alignment is not exceptionally poor, no significant angular dispersion can develop in the vertical plane: in case that the grooves of the grating

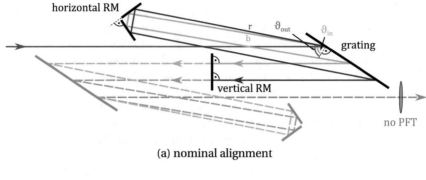

(a) nominal alignment

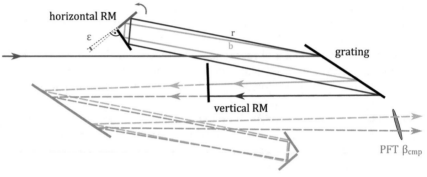

(b) slightly misaligned horizontal roof mirror

Fig. 6.10 Sketch of the beam path in a **a** perfectly aligned and **b** slightly misaligned folded grating compressor. View is from top, the beam path is shown for two colors, r(ed) and b(lue). Dashed and pale elements correspond to the second pass through the compressor which is displayed here separately for a better visibility. In case of a slightly misaligned—i.e. non-orthogonal—horizontal roof mirror (RM) the pulse front of the output pulses is tilted

are not exactly perpendicular to the table, an angular dispersion component in the vertical dimension is created. However, the vertical RM (= end-mirror) inverts this dispersion and therefore it is compensated by the second pass on the way back, with just a remaining spatial chirp.

In contrast, angular dispersion in the horizontal plane introduced by a non-orthogonal horizontal RM does not cancel but doubles. A schematic illustration of the beam paths is shown in Fig. 6.10. Note that when tilting the horizontal RM, the vertical RM has to be rotated to make sure that the direction of the beam exiting the compressor stays unchanged.

Similar to Eq. (6.10), a PFT angle β_{cmp} at the output of the compressor is created by tilting one mirror of the horizontal RM as indicated in Fig. 6.10 by the angle ε:

$$\varepsilon \approx \frac{1}{4} \frac{\cos \vartheta_{in}}{\tan \vartheta_{out}} \frac{d}{\lambda_0} \beta_{cmp} \tag{6.13}$$

where ϑ_{in} and ϑ_{out} are the angles of incident and diffracted beam relative to the grating normal, d is again the grating constant and λ_0 is the central wavelength. The additional factor of $1/4$ originates from the double-pass through the compressor and the angle-doubling by reflection of the beam on the tilted mirror. The required PFT angle β_{cmp} at the compressor can be calculated from the PFT angle β_{SHG} at the SHG crystal (cf. Eq. (6.8)) by taking into account the magnification M_{cmp} of the intermediate imaging telescope:

$$\beta_{cmp} = M_{cmp}\,\beta_{SHG} \approx M_{cmp}\,M_{SHG}\,n_{LBO,515\,nm}\,\alpha_{int} \tag{6.14}$$

As the PFT of the fundamental pulses directly transfers to the frequency-doubled pulses in the SHG process, no further conversion is required. The tolerance for the tilt angle of the horizontal RM is given by:

$$\Delta\varepsilon = \frac{1}{4}\frac{\cos\vartheta_{in}}{\tan\vartheta_{out}}\frac{d}{\lambda_0}\,\Delta\beta_{cmp} = \frac{1}{4}\frac{\cos\vartheta_{in}}{\tan\vartheta_{out}}\frac{d}{\lambda_0}\,M_{cmp}\,M_{SHG}\,n_{LBO,515\,nm}\,\frac{30\,\mu m}{D_{OPA}}$$
$$\tag{6.15}$$

For the 9 J fundamental beam that is later used to pump the 3rd OPA stage, the planned parameters are: $\vartheta_{in} = 60°$, $\vartheta_{out}(\lambda_0) = 67.8°$, $d = 1/1740$ lines/mm, $\lambda_0 = 1030$ nm, $M_{cmp,3} = 0.8$, $M_{SHG,3} = 1$ and $D_{OPA,3} = 65$ mm. This yields an optimal tilt angle of $\varepsilon_3 = 0.0401(10)°$ for the horizontal RM. Although setting the mirror angle to this value within the permissible tolerance requires a highly accurate alignment, the large size of the horizontal RM substrates ($\sim$200 mm) creates a lever that is expected to make this precision accessible.

The proposed adjustment of the PFT in the fundamental beam by a controlled misalignment of the pump compressor has a few consequences that need to be considered:

First, by propagation inside the pump compressor chamber the angularly dispersed beam aggregates a *spatial chirp*. However, at an angular chirp of

$$C_a \approx \frac{\beta_{cmp}}{\lambda_0} \approx 24\,\frac{\mu rad}{nm} \approx 24\,\frac{\mu m}{nm\,m} \tag{6.16}$$

and a propagation distance L of only few meters, the lateral wavelength separation of $C_s = L\,\Delta\lambda\,C_a$ is (at a bandwidth of $\Delta\lambda \sim 3$ nm) less than 1 mm and thus small compared to the beam diameter $D_{cmp} = 65$ mm. Any further increase of this spatial chirp after the compressor chamber on the other hand is compensated by the already described telescope imaging system.

A second aspect is the one-dimensionally enlarged focal spot size (due to angular dispersion) in the telescope between compressor and SHG crystal that might affect the planned *spatial filtering* of the beam with a pinhole. At about 75 μrad, however, the angle spread is smaller than the 200 μrad diameter (in angular space) of the pinhole that is currently employed in the spatial filter of the 1 J system. This pinhole size has experimentally been found to be optimal by choosing the diameter as small as possible for efficient spatial filtering while still keeping energy losses lower than a few percent. As wavefront distortions and pointing will presumably require a similar pinhole diameter for the 9 J beam line, the angular dispersion from pulse front tilting should therefore not be the limiting factor.

Finally, a decrease in **SHG efficiency** due to the angular chirp is not expected as the range of angles is $\pm 37\,\mu$rad and hence about two orders of magnitude smaller than the acceptance angle of the SHG crystal at ± 4.3 mrad (cf. Fig. 5.4a).

PFT Control for the First and Second OPA Stage

As already mentioned, in the planned setup the 9 J and 1 J pump pulses will have separate beam paths inside the compressor with separate horizontal roof mirrors. This important fact allows an adjustment of the pump PFT for the second OPA stage *independently* from the optimization of the third stage that has been discussed so far. With the planned telescope magnifications $M_{cmp,2} = 1.4$ and $M_{SHG,2} = 0.56$, the target tilt angle of the horizontal RM is accordingly $\varepsilon_2 = 0.0391\,(31)°.$[9]

In the current (as well as in the planned) system, the frequency-doubled 1 J-pulses are split for the first and second OPA stage only after SHG. Therefore, the PFT that is introduced by compressor adjustment in the pump for the second stage will also affect the pump for the first stage. However, since the telescope magnifications M_{SHG} of the two stages are different, the generated PFT of $\beta_{int} = 4.2°$ at the first stage does not match the target angle $\alpha_{int} = 1.1°$. Even though the PFT tolerance is at $\Delta\beta_{int,1} = 0.34°$ comparably large due to the small beam size at this stage, the deviation $|\gamma_{int}| = |\alpha_{int} - \beta_{int}| = 3.1°$ exceeds this value significantly and hence requires a compensation.

For this reason, a pair of transmission gratings for 515 nm was purchased from Fraunhofer IOF ($75 \times 85 \times 3$ mm, 65×75 mm grating area, 2000 lines/mm, $\vartheta_L = 31.0°$, eff. $>95\%$). Placed behind the beam splitter in the optical path of the pump for the first OPA stage, the intensity is very low at about 1.7 mJ/cm^2 and hence the risk of laser-induced damage as well as B-integral can be neglected. This allowed us to use thicker grating substrates with a good flatness (peak to valley $\sim 4\,\mu$m). Furthermore, in contrast to the stitched gratings from LightSmyth, the pattern of the IOF gratings is written as a whole (electron beam lithography) and is hence smooth.

Tilting these gratings relative to each other by $\varphi = 1.01(11)°$ will compensate the PFT introduced by the pump compressor and provide the required angles $\beta_{SHG,1} = 0.26°$ and $\beta_{int,1} = 1.1°$. The complete proposed setup is shown schematically in Fig. 6.11 and the corresponding parameters are once more listed in Table 6.3.

---

[9]One might notice that within the tolerance, ε_2 is equal to ε_3. This is due to the fact that the summed up telescope magnifications planned for the 1 J and 9 J beams from the compressor to the OPA stages are quasi-identical. As the setup, however, might change in the future, an independent optimization is nevertheless essential.

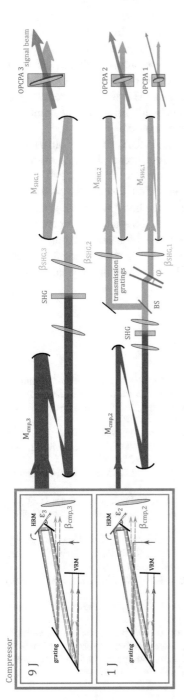

Fig. 6.11 Sketch of the suggested setup to control the pulse front tilt of pump beams for all three OPCPA stages. Note that in the real compressor setup the 9 and 1 J beams share one large grating with separate roof mirrors on different heights

Table 6.3 Configuration parameters of the suggested setup. D are FWHM beam diameters, β are PFT angles, M are telescope magnifications, ε are tilt angles for the horizontal roof mirror in the pump compressor and φ is the tilt angle between the transmission gratings. Tolerance values define the range within which the corresponding PFT mismatch in the OPCPA stages is still acceptable. For details see main text

OPCPA stage	D_{cmp} (mm)	$\varepsilon \pm \Delta\varepsilon$	$\beta_{\mathrm{cmp}} \pm \Delta\beta_{\mathrm{cmp}}$	M_{cmp}	D_{SHG} (mm)	$\varphi \pm \Delta\varphi$	$\beta_{\mathrm{SHG}} \pm \Delta\beta_{\mathrm{SHG}}$	M_{SHG}	D_{OPA} (mm)	$\beta_{\mathrm{int}} \pm \Delta\beta_{\mathrm{int}}$
3	81	$0.0401 \pm 0.0010°$	$1.413 \pm 0.036°$	0.8	65	–	$1.766 \pm 0.045°$	1	65	$1.100 \pm 0.028°$
2	25	$0.0391 \pm 0.0031°$	$1.38 \pm 0.11°$	1.4	35	–	$0.987 \pm 0.079°$	0.56	20	$1.100 \pm 0.088°$
1	— "— —	— "— —	— "— —	— "— —	— "— —	$1.01 \pm 0.11°$	$0.256 \pm 0.079°$	0.145	5.1	$1.10 \pm 0.34°$

Summary

In summary, a scheme was suggested that matches the pulse fronts of signal and pump beams in the three individual OPA stages by tilting the pump pulse fronts in accordance with the non-collinear angle. For the second and third OPA stage this tilt will be introduced by a controlled misalignment of the pump compressor. As the pump beam for the first OPA stage is affected by this measure as well, its pulse front tilt will be corrected by an additionally implemented pair of transmission gratings. For a potential fourth OPA stage the pump pulses of the third stage are recycled, hence pulse front matching with the signal is automatically fulfilled if the number of mirror reflections between the two stages is properly chosen.

At the time this thesis is written, a rotating periscope in the vacuum system between pump compressor and SHG crystal (cf. Fig. 3.9) still prevents the implementation of the suggested solution: since in the compressor only a horizontal PFT can be generated which is rotated by the periscope to a vertical PFT, so far the pulse fronts of pump and signal (intersecting at the OPA stages in the horizontal plane) cannot be matched. In the currently ongoing reconstruction of the PFS system in preparation for the upgrade, however, this shortcoming will be removed by a new and non-rotating vacuum connection for both 1 and 9 J beam. Furthermore, a new motorization of the horizontal and vertical RMs in the compressor will allow the required adjustment of the PFT even if the system is evacuated.

References

1. C. Wandt, S. Klingebiel, S. Keppler, M. Hornung, C. Skrobol, A. Kessel, S.A. Trushin, Z. Major, J. Hein, M.C. Kaluza, F. Krausz, S. Karsch, Development of a Joule-class Yb:YAG amplifier and its implementation in a CPA system generating 1 TW pulses. Laser Photonic Rev. **881**, 875–881 (2014). https://doi.org/10.1002/lpor.201400040
2. M. Schultze, T. Binhammer, G. Palmer, M. Emons, T. Lang, U. Morgner, Multi-μJ, CEP-stabilized, twocycle pulses from an OPCPA system with up to 500 kHz repetition rate. Opt. Express **18**, 27291–27297 (2010). https://doi.org/10.1364/OE.18.027291
3. H. Fattahi, *Third-generation femtosecond technology*, Ludwig-Maximilians-Universität München, Ph.D. thesis, 2015
4. J. Ahrens, O. Prochnow, T. Binhammer, T. Lang, B. Schulz, M. Frede, U. Morgner, Multipass OPCPA system at 100 kHz pumped by a CPA-free solid-state amplifier. Opt. Express **24**, 8074 (2016). https://doi.org/10.1364/OE.24.008074
5. Z. Major, S.A. Trushin, I. Ahmad, M. Siebold, C. Wandt, S. Klingebiel, T.-J. Wang, J.A. FÜlöp, A. Henig, S. Kruber, R. Weingartner, A. Popp, J. Osterhoff, R. HOrlein, J. Hein, V. Pervak, A. Apolonski, F. Krausz, S. Karsch, Basic concepts and current status of the petawatt field synthesizer-a new approach to ultrahigh field generation. Rev. Laser Eng. **37** 431–436 (2009). https://doi.org/10.2184/lsj.37.431
6. C. Skrobol, *High-Intensity, Picosecond-Pumped, Few-CycleOPCPA* (Ludwig-Maximilians-Universität München, PhDthesis, 2014)
7. A.V. Smith, B.T. Do, Bulk and surface laser damage of silica by picosecond and nanosecond pulses at 1064 nm. Appl. Opt. **47**, 4812–4832 (2008). https://doi.org/10.1364/AO.47.004812
8. G. Pretzler, A. Kasper, K. Witte, Angular chirp and tilted light pulses in CPA lasers. Appl. Phys. B: Lasers Opt. **70**, 1–9 (2000). https://doi.org/10.1007/s003400050001

9. J.A. FÜlöp, Z. Major, A. Henig, S. Kruber, R. Weingartner, T. Clausnitzer, E.-B. Kley, A. TUnnermann, V. Pervak, A. Apolonski, J. Osterhoff, R. HÖrlein, F. Krausz, S. Karsch, Short-pulse optical parametric chirped-pulse amplification for the generation of high-power few-cycle pulses. New J. Phys. **9**, 438–438, (2007). https://doi.org/10.1088/1367-2630/9/12/438

10. C. Zhou, T. Seki, T. Kitamura, Y. Kuramoto, T. Sukegawa, N. Ishii, T. Kanai, J. Itatani, Y. Kobayashi, S. Watanabe, Wavefront analysis of high-efficiency, large-scale, thin transmission gratings. Springer Proc. Phys. **162**, 779–782 (2015). https://doi.org/10.1007/978-3-319-13242-6_191

11. C.M. Greiner, D. Iazikov, T.W. Mossberg, Diffraction-limited performance of flat-substrate reflective imaging gratings patterned by DUV photolithography. Opt. Express **14**, 11952–11957 (2006). https://doi.org/10.1364/OE.14.011952

12. A. Popp, J. Vieira, J. Osterhoff, Z. Major, R. HOrlein, M. Fuchs, R. Weingartner, T.P. Rowlands-Rees, M. Marti, R.A. Fonseca, S.F. Martins, L.O. Silva, S.M. Hooker, F. Krausz, F. GrUner, S. Karsch, All-optical steering of laser-wakefield-accelerated electron beams. Phys. Rev. Lett. **105**, 1–4 (2010). https://doi.org/10.1103/PhysRevLett.105.215001

13. M. Schnell, A. SÄvert, I. Uschmann, M. Reuter, M. Nicolai, T. KÄmpfer, B. Landgraf, O. JÄckel, O. Jansen, A. Pukhov, M.C. Kaluza, C. Spielmann, Optical control of hard X-ray polarization by electron injection in a laser wakefield accelerator. Nat. Commun. **4**, 2421 (2013). https://doi.org/10.1038/ncomms3421

14. S. Klingebiel, Picosecond pump dispersion management and jitter stabilization in a petawatt-scale few-cycle OPCPA system, Ph.D. thesis, Ludwig-Maximilians-Universität München, 2013

15. M. Lai, S.T. Lai, C. Swinger, Single-grating laser pulse stretcher and compressor. Appl. Opt. **33**, 6985–6987 (1994). https://doi.org/10.1364/AO.33.006985

Chapter 7
Summary and Outlook

7.1 Summary

In this work, the recent progress of the Petawatt Field Synthesizer has been described, a system that is designed to generate few-cycle light pulses at highly relativistic intensities by broadband OPCPA with picosecond pumping. On the route towards the experimental realization of this system with its unrivaled target parameters, important milestones have been reached.

Motivated by the poor performance of the prior method of broadband seed generation, three alternative schemes have been developed and tested: The idler generation scheme in principle provides high pulse energies ($10\,\mu$J) and the required broad spectral bandwidth (680–1400 nm) but proved unusable due to an uncompensatable angular chirp in the generated pulses. As a second scheme, the previous seed generation setup based on two hollow-core fibers was improved by adding an XPW-stage to temporally gate the spectrally broadened pulses, yielding $4\,\mu$J pulse energy and a very smooth broadband spectrum. These pulses were used on a daily basis as the seed for OPCPA experiments. The third tested scheme—based on cascaded nonlinear processes—promises significantly higher pulse energies (tens of μJ) and is suggested as a future seed source.

In a low-energy OPCPA campaign, the XPW seed pulses were amplified in a vacuum-based OPA system with two stages (LBO) to pulse energies of up to 10 mJ. Test compression of the resulting broadband pulses (700–1300 nm) with chirped mirrors in air yielded a pulse duration of 7 fs and demonstrated the suitability of the XPW pulses as a seed for the OPA chain.

In preparation for the high-energy OPCPA campaign, several improvements were made to the pump system: the replacement of the master oscillator allowed direct seeding of the Yb-based pump chain at 1030 nm, the last amplifier was replaced by a more reliable and more energetic version and the pump compressor was rebuilt in vacuum to improve stability and support higher pulse energies. Furthermore, the SHG efficiency of the compressed pump pulses was significantly increased from 45

© Springer International Publishing AG, part of Springer Nature 2018
A. Kessel, *Generation and Parametric Amplification of Few-Cycle
Light Pulses at Relativistic Intensities*, Springer Theses,
https://doi.org/10.1007/978-3-319-92843-2_7

to 55% by temporal gating of the fundamental pulses with a fast-switching Pockels cell that suppressed a pronounced ASE pedestal.

Pumping the two OPA stages at 10 Hz with 14 and 400 mJ respective pump energy resulted in an amplified signal pulse energy of 1 mJ after the first and 53 mJ after the second stage. Staying inside the vacuum system, the amplified pulses were temporally compressed by 14 reflections on chirped mirrors. In combination with adaptive dispersion control of the seed pulses before amplification, a pulse duration of 6.4 fs (2.2-cycles) was achieved. In total this yielded a peak power of up to 4.9 TW after compression and a peak intensity of up to 4.5×10^{19} W/cm^2 after focusing.

The measurement of the temporal contrast revealed a ratio of at least 10^{11} between the main peak and the background—an excellent value as expected from OPA systems due to the instantaneous transfer of energy from pump to signal beam. Moreover, it could be shown that owing to the sub-picosecond pump pulse duration, there is no pedestal around the main peak (at least not longer than $\sim$1 ps), a crucial advantage of the PFS over OPCPA systems with longer pump pulse durations (e.g. [1, 2]) especially for the key application of surface high-harmonic generation.

Furthermore, preparations were made for an upgrade of the PFS system that is expected to boost the output power in a third and potentially fourth OPA stage by more than an order of magnitude towards the 100 TW regime. The progress in crystal manufacturing in the past years allows us to use 80 mm LBO crystals for the new OPA stages and for frequency-doubling of the pump pulses instead of DKDP crystals, as originally planned. To determine the optimum crystal parameters for our purposes, we conducted a series of test measurements that suggest a thickness of 1.5 mm for the pump SHG crystal at a fluence of 0.16 J/cm^2 (fundamental beam). For the OPA crystals, the preliminary test results revealed for different crystal thicknesses a trade-off between amplified pulse energy, spectral smoothness and stability. The final decision will be based on the outcome of a planned second measurement series.

Finally, it was shown that due to the short pulse durations and the large beam diameters in the upgraded system, matching the pulse fronts of pump and signal pulses at the OPA stages is indispensable to guarantee a good spatio-temporal overlap. To this end, two potential schemes were analyzed that tilt the pulse fronts of the pump pulses. The first scheme—comprising a pair of transmission gratings—turned out to be only usable for low-intensity pulses because of the low damage threshold of the gratings. Hence, a second scheme was considered where the PFT is created by a controlled misalignment of the pump compressor. Based on a quantitative analysis, a combination of both schemes was proposed to independently match the pulse fronts of pump and signal at all three (or even four) OPA stages of the upgraded system.

7.2 High Peak-Power Systems Worldwide

As roughly ten years have passed now since the first design of the PFS concept, it is a good idea to take a step back, have a look at the current status of the project and compare its present as well as expected future performance to other high-energy systems that have been built or designed in the meanwhile.

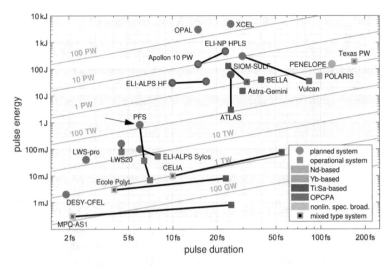

Fig. 7.1 Selection of worldwide light sources. Connected markers correspond to identical or closely related systems. DESY-CFEL [8], MPQ-AS1 [9], LWS-pro [10], Ecole Polytechnique [11], LWS20 [12], ELI-ALPS Sylos [2], CELIA [13], ELI-ALPS HF [5], Apollon 10 PW [6], OPAL [14], ATLAS [15], ELI-NP HPLS [16], SIOM-SULF [4, 7], XCEL [17], Astra-Gemini [18], BELLA [19], Vulcan [20, 21], POLARIS [3], PENELOPE [22], Texas PW [23]. It should be noted that the plot does not provide any information about the repetition rate. As this quantity varies for the examined systems from few shots per day to multi-kHz the choice of the optimum system for a specific application depends very much on whether the peak or the average power is more important. In this respect, recently developed systems with μJ pulse energy but very high repetition rates of 0.3–1.27 MHz are worth mentioning [24, 25]

Figure 7.1 shows a selection of light sources including the PFS system with the parameters achieved in this work and expected from the upgrade. In the examined temporal region (1.5–200 fs) different concepts can be identified: at comparably long pulse durations there are a few light sources around the 1 PW-level which are based on Nd- or Yb-doped gain media such as the all-diode-pumped POLARIS system in Jena [3]. The majority of existing and planned systems at or above 1 PW, however, is represented by Ti:Sa-based systems in the 10–50 fs region. Notably, at 5 PW the highest-peak-power system existing today is also based on Ti:Sa as a gain medium [4]. Owing to the high contrast achievable by OPCPA, it has become popular to combine an OPCPA frontend with one or multiple successive conventional laser amplifiers that boost the pulse energy to the desired value [5, 6]. Experiments with the inverted scheme (Ti:Sa frontend, OPCPA booster) have also been reported [7]. Regarding the pulse duration, however, these schemes are restricted to the region >10 fs. This applies also for the ambitious OPAL and XCEL projects that seek to bring OPCPA at moderate bandwidth to the kilojoule-level.

The few- and single-cycle domain 2–10 fs can be accessed by either broadband OPCPA or by nonlinear spectral broadening of conventionally generated laser pulses. While the latter technique has been demonstrated so far only up to the 1 TW level as shown in the plot, theoretical studies suggest to apply it also on a Petawatt scale [26, 27]. Depending on the experimental success of the proposed schemes, they might provide an attractive way towards ultrashort, high-intensity light pulses.

For the time being, pure OPCPA systems promise the best performance in the sub-10 fs regime. Currently, the most powerful few-cycle light source at ∼20 TW is the LWS20 (Light Wave Synthesizer) [12], a two-color-pumped (2ω and 3ω of Nd:YAG) OPCPA system operating at 10 Hz that has recently moved from the MPQ to Umeå University. Other systems trying to reach a comparable peak power (at higher repetition rates) are the LWS-pro at the Laboratory for Extreme Photonics (LEX) [10] as well as the Sylos system as part of the ELI-ALPS (Extreme Light Infrastructure—Attosecond Light Pulse Source) project in Hungary [2].

The PFS, at the parameters achieved in this work, is among the leading systems in the sub-10 fs category, as can be seen from the plot. The upgrade that is currently under construction is expected to boost the PFS pulse energy into a region, no few-cycle laser has reached so far, enabling experiments with unprecedented pulse parameters. In that sense the long development time of the PFS will hopefully still pay off.

7.3   Current Works and Outlook

Using the OPCPA system as described in Chap. 5, first HHG experiments on solid surfaces have been conducted by Olga Lysov and Vyacheslav Leshchenko (both from our group). To this end the compressed signal pulses were magnified in a reflective telescope and guided to the target chamber where a f/1.2 45° off-axis parabolic mirror focused them onto a fused silica substrate. By motorized rotation of the target, it was made sure that every shot hits a fresh spot on the surface. For various reasons, in this campaign the pulse energy on target was only ∼10 mJ at a pulse duration of ∼6.5 fs. The focal spot diameter was measured to be 1.8 μm FWHM with about 70% of the total energy within the central peak. In total this yields a peak intensity of 2.1×10^{19} W/cm^2 corresponding to a normalized vector potential of $a_0 \approx 3$.

Figure 7.2 shows the spectrum of the first high-harmonic pulses generated with the ROM technique described in the introduction. The displayed data were acquired by removing the fundamental beam with an Al-filter and guiding the transmitted harmonic pulses to an XUV-spectrometer consisting of a grating and a CCD. The measured spectral modulations in Fig. 7.2 correspond to the 17th to 30th harmonic, where the decrease of intensity at the left (long wavelengths) is only due to the spectral sensitivity of the diagnostics. Having conducted this proof-of-principle experiment, it is planned to examine the spectral cut-off at the short wave end and shift it as far as possible into the XUV by further optimization of the driving pulses, i.e. bandwidth, temporal shape and energy on target. Furthermore, by tagging the CEP, the dependence of the XUV spectra on the absolute phase of the generating pulses

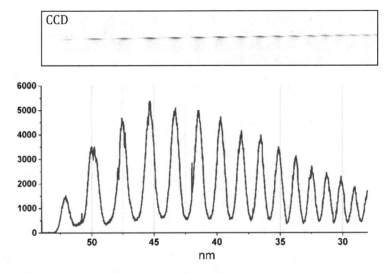

Fig. 7.2 First XUV spectrum generated with the PFS by high-harmonic generation on solid surfaces. The picture in the upper panel shows the raw CCD data from the spectrometer, the plot in the lower panel displays the pictures' line-out after vertical binning (Courtesy of Olga Lysov and Vyacheslav Leshchenko)

will be analyzed. Finally, it is planned to temporally characterize the high-harmonic radiation to verify the generation of isolated attosecond pulses.

At the moment the preparations for the PFS upgrade are under way and first changes to the setup have already been made: the 1 J-line in the pump compressor has been rebuild to give space for the compression of the future 9 J pulses and additionally the vacuum beam line and spatial filter between compressor and SHG has been replaced by a new version that allows us to apply the suggested method for pulse front matching.

Depending on the pace of progress in the pump amplifier development, the first tests with the upgraded system will still be conducted at the MPQ or later at the nearby Centre for Advanced Laser Applications (CALA) [28] to which the PFS will move by the end of this year (2017). Together with its sister system—the PFS-pro that aims at lower pulse energies but higher repetitions rates (100 mJ at 10 kHz)—it will provide light pulses for HHG on solid surfaces, ion-acceleration [29], and Thomson-scattering experiments.

References

1. J.M. Mikhailova, A. Buck, A. Borot, K. Schmid, C. Sears, G.D. Tsakiris, F. Krausz, L. Veisz, Ultra-highcontrast few-cycle pulses for multipetawatt-class laser technology. Opt. Lett. **36**, 3145 (2011). https://doi.org/10.1364/OL.36.003145

2. R. Budri unas,T. Stanislauskas, J. Adamonis, A. AleknaviCius, G. Veitas, D.Gadonas, S. Balickas, A. Michailo-vas, A. VaranaviCius, 53Waverage power CEP-stabilized OPCPA system delivering 55 TWfew cycle pulses at 1 kHz repetition rate. Opt. Express **25**, 5797 (2017). https://doi.org/10.1364/OE.25.005797

3. M. Hornung, H. Liebetrau, S. Keppler, A. Kessler, M. Hellwing, F. Schorcht, G.A. Becker, M. Reuter, J. Polz, J. KOrner, J. Hein, M.C. Kaluza, 54 J pulses with 18 nm bandwidth from a diode-pumped chirpedpulse amplification laser system. Opt. Lett. **41**, 5413 (2016). https://doi.org/10.1364/OL.41.005413

4. Z. Gan, L. Yu, S. Li, C. Wang, X. Liang, Y. Liu, W. Li, Z. Guo, Z. Fan, X. Yuan, L. Xu, Z. Liu, Y. Xu, J. Lu, H. Lu, D. Yin, Y. Leng, R. Li, Z. Xu, 200 J high efficiency Ti : sapphire chirped pulse amplifier pumped by temporal dual- pulse. Opt. Express **25**, 5169–5178 (2017). https://doi.org/10.1364/OE.25.005169

5. M. Aeschlimann et al., ELI-ALPS Scientific Case, http://www.eli-alps.hu/sites/default/files/The_Scientific_Case_of_ELI-ALPS_2015_0.pdf. Accessed 17 Mar 2017

6. D.N. Papadopoulos, J. Zou, C.L. Blanc, G. Ch, A. Beluze, N. Lebas, P. Monot, F. Mathieu, P. Audebert, The Apollon 10PWlaser: experimental and theoretical investigation of the temporal characteristics. High Power Laser Sci. Eng. **4**, 1–7 (2016). https://doi.org/10.1017/hpl.2016.34

7. L. Yu, Z. Xu, X. Liang, L. Xu, W. Li, C. Peng, Z. Hu, C. Wang, X. Lu, Y. Chu, Z. Gan, X. Liu, Y. Liu, X. Wang, H. Lu, D. Yin, Y. Leng, R. Li, Z. Xu, Optimization for high-energy and high-efficiency broadband optical parametric chirped-pulse amplification inLBOnear 800 nm. Opt. Lett. **40**, 3412 (2015). https://doi.org/10.1364/OL.40.003412

8. F.X. KÄrtner, O. MÜcke, G. Cirmi, S. Fang, S.-H. Chia, C. Manzoni, P. Farinello, G. Cerullo, High energy sub-cycle optical waveform synthesizer, Advanced Solid-State Lasers Congress, AW2A.1 (2013). https://doi.org/10.1364/ASSL.2013.AW2A.1

9. A. Wirth, M.T. Hassan, I. Grguras, J. Gagnon, A. Moulet, T.T. Luu, S. Pabst, R. Santra, Z.A. Alahmed, A.M. Azzeer, V.S. Yakovlev, V. Pervak, F. Krausz, E. Goulielmakis, Synthesized light transients. Science **334**, 195–200 (2011). https://doi.org/10.1126/science.1210268

10. H. Fattahi, Third-generation femtosecond technology, Ph.D. thesis, Ludwig-Maximilians-Universitat nchen, 2015

11. F. BÖhle, M. Kretschmar, A. Jullien, M. Kovacs, M. Miranda, R. Romero, H. Crespo, P. Simon, R. Lopez- Martens, T. Nagy, Generation of 3-mJ, 4-fs CEP-Stable pulses by long stretched flexible hollow fibers. Research in Optical Sciences: Postdeadline Papers, HW5C.2 (2014). https://doi.org/10.1364/HILAS.2014.HW5C.2

12. L. Veisz, D. Rivas, G. Marcus, X. Gu, D. Cardenas, J. Xu, J. Mikhailova, A. Buck, T. Wittmann, C.M.S. Sears, D. Herrmann, O. Razskazovskaya, V. Pervak, F. Krausz, Multi-10-TWsub-5-fs optical parametric synthesizer, in *2014 IEEE Photonics Conference*, vol. 163 (2014), pp. 510–511. https://doi.org/10.1109/IPCon.2014.6995473

13. O. Hort, A. Dubrouil, A. Cabasse, S. Petit, E. MEvel, D. Descamps, E. Constant, Postcompression of high-energy terawatt-level femtosecond pulses and application to high-order harmonic generation. J. Opt. Soc. Am. B **32**, 1055 (2015). https://doi.org/10.1364/JOSAB.32.001055

14. J. D. Zuegel, S.-W. Bahk, I. A. Begishev, J. Bromage, C. Dorrer, A. V. Okishev, J.B. Oliver, Status of high-energy OPCPA at LLE and future prospects, in *CLEO 2014, JTh4L*, vol. 4 (2014). https://doi.org/10.1364/CLEO_AT.2014.JTh4L.4

15. The Advanced Titanium-Sapphire Laser (ATLAS), http://www.cala-laser.de/en/instruments/light-sources.html. Accessed 17 Mar 2017

16. F. Lureau, S. Laux, O. Casagrande, O. Chalus, A. Pellegrina, G. Matras, C. Radier, G. Rey, S. Ricaud, S. Herriot, P. Jougla, M. Charbonneau, P. Duvochelle, C. Simon-Boisson, Latest results of 10 petawatt laser beamline for ELI nuclear physics infrastructure, in *Proceedings of the SPIE*, vol. 9726 (2016). https://doi.org/10.1117/12.2213067

17. XCEL project website, http://www.xcels.iapras.ru/. Accessed 12 Apr 2017

18. B. Parry, C. Hooker, Y. Tang, Dual beam operation of the Astra-Gemini high power laser and upgrades to the Ti:Sapphire amplifiers, in *2012 Conference on Lasers and Electro-Optics, CLEO 2012* (2012), pp. 1–2

19. W.P. Leemans, J. Daniels, A. Deshmukh, A.J. Gonsalves, A. Magana, H.-S. Mao, D.E. Mittel-berger, K. Naka-Mura, J. R. Riley, D. Syversrud, C. TOth, N. Ybarrolaza, BELLA Laser and operations, in *Proceedings of PAC* (2013), pp. 1097–1100
20. O.V. Chekhlov, J.L. Collier, I.N. Ross, P.K. Bates, M. Notley, C. Hernandez-Gomez, W. Shaikh, C. N. Dan-son, D. Neely, P. Matousek, S. Hancock, L. Cardoso, 35 J broadband femtosecond optical parametric chirped pulse amplification system. Opt. Lett. **31**, 3665 (2006). https://doi.org/10.1364/OL.31.003665
21. C. Hernandez-Gomez, S.P. Blake, O. Chekhlov, R.J. Clarke, A.M. Dunne, M. Galimberti, S. Hancock, R. Heathcote, P. Holligan, A. Lyachev, P. Matousek, I.O. Musgrave, D. Neely, P.A. Norreys, I. Ross, Y. Tang, T.B. Winstone, B.E. Wyborn, J. Collier, The Vulcan 10 PW project. J. Phys. Conf. Ser. **244**, 032006 (2010). https://doi.org/10.1088/1742-6596/244/3/032006
22. M. Siebold, F. Roeser, M. Loeser, D. Albach, U. Schramm, PEnELOPE: a high peak-power diode-pumped laser system for laser-plasma experiments. Proc. SPIE **8780**, 878005 (2013). https://doi.org/10.1117/12.2017522
23. E.W. Gaul, M. Martinez, J. Blakeney, A. Jochmann, M. Ringuette, D. Hammond, T. Borger, R. Escamilla, S. Douglas, W. Henderson, G. Dyer, A. Erlandson, R. Cross, J. Caird, C. Ebbers, T. Ditmire, Demonstration of a 1.1 petawatt laser based on a hybrid optical parametric chirped pulse amplification/mixed Nd:glass amplifier. Appl. Opt. **49**, 1676–1681 (2010). https://doi.org/10.1364/AO.49.001676
24. M. Schultze, S. Prinz, M. Haefner, C.Y. Teisset, R. Bessing, T. Metzger, High-power 300-kHz OPCPA system generating CEP-stable few-cycle pulses (2015), pp. 5–6. https://doi.org/10.1364/CLEO_SI.2015.SF1M.6
25. S. HÄdrich, M. Kienel, M. MÜller, A. Klenke, J. Rothhardt, R. Klas, T. Gottschall, T. Eidam, A. Drozdy, P. JOjArt, Z. VArallyay, E. Cormier, K. Osvay, A. TUnnermann, J. Limpert, Energetic sub-2-cycle laser with 216Waverage power. Opt. Lett. **41**, 4332–4335 (2016). https://doi.org/10.1364/OL.41.004332
26. S. Mironov, E. Khazanov, G. Mourou, Pulse shortening and ICR enhancement for PW-class lasers. Specialty Optical Fibers, JTu3A.24 (2014)
27. G. Mourou, S. Mironov, E. Khazanov, A. Sergeev, Single cycle thin film compressor opening the door to Zeptosecond-Exawatt physics. Eur. Phys. J. Special Top. **223**, 1181–1188 (2014). https://doi.org/10.1140/epjst/e2014-02171-5
28. CALA project website, http://www.cala-laser.de/. Accessed 17 Mar 2017
29. M.L. Zhou, X.Q. Yan, G. Mourou, J.A. Wheeler, J.H. Bin, J. Schreiber, T. Tajima, Proton acceleration by single-cycle laser pulses offers a novel monoenergetic and stable operating regime. Phys. Plasmas **23**, 1–6 (2016). https://doi.org/10.1063/1.4947544

Appendix A
Supplementary Calculations and Experiments

A.1 Spectral Conversion Between Wavelength and Frequency Space

Due to the reciprocal relation between wavelength and angular frequency

$$\omega = \frac{2\pi c}{\lambda} \tag{A.1}$$

the conversion of a laser pulse spectrum from one space to the other is not a linear transformation but is defined by

$$S_\omega(\omega) = S_\omega(2\pi c/\lambda) = S_\lambda(\lambda) \frac{\lambda^2}{2\pi c} \tag{A.2}$$

Neglecting this rule by assuming that $S_\omega(\omega) = S_\lambda(\lambda)$ is a frequently encountered mistake that modifies the energy density in each spectral region and leads for example to a wrong calculation of the Fourier limited pulse duration.

A.2 Nonlinear Crystals

A.2.1 Contracted Susceptibility Tensor Notations

Instead of the full n-dimensional nonlinear tensors

$$d_{ijk} = \frac{1}{2}\chi^{(2)}_{ijk} \qquad\qquad c_{ijkl} = \chi^{(3)}_{ijkl} \tag{A.3}$$

© Springer International Publishing AG, part of Springer Nature 2018
A. Kessel, *Generation and Parametric Amplification of Few-Cycle Light Pulses at Relativistic Intensities*, Springer Theses,
https://doi.org/10.1007/978-3-319-92843-2

Table A.1 Corresponding indices in original and contracted notation of the second- and third-order nonlinear tensors

$d_{il} \leftrightarrow d_{ijk}$:	l	1	2	3	4	5	6
	jk	11	22	33	23,32	13,31	12,21

$c_{im} \leftrightarrow c_{ijkl}$:	m	1	2	3	4	5	6
	jkl	111	222	333	233,323,332	223,232,322	133,313,331

	7	8	9	0
	113,131,311	122,212,221	112,121,211	123,132,213,231,312,321

often the contracted notations d_{il} and c_{im} are used that take advantage of crystal symmetries [1]. The corresponding indices in the original and contracted notation are listed in Table A.1.

A.2.2 Coordinate Transformation of the Susceptibility Tensor

If a nonlinear crystal is rotated with respect to the beam coordinate system, one can use the rotation matrix T

$$T(\varphi, \vartheta) = \begin{pmatrix} -\cos\varphi \cos\vartheta & -\sin\varphi \cos\vartheta & \sin\vartheta \\ \sin\varphi & -\cos\varphi & 0 \\ \cos\varphi \sin\vartheta & \sin\varphi \sin\vartheta & \cos\vartheta \end{pmatrix} \quad (A.4)$$

to calculate its effective nonlinear susceptibility tensors. For this purpose, one should first consider the transformation of an electric field E defined in crystal coordinates (X,Y,Z) into the field E' in beam coordinates (x,y,z):

$$\begin{pmatrix} E_x \\ E_y \\ E_z \end{pmatrix} = E' = T(\varphi, \vartheta) \cdot E = T(\varphi, \vartheta) \cdot \begin{pmatrix} E_X \\ E_Y \\ E_Z \end{pmatrix} \quad (A.5)$$

and vice versa:

$$E = T^{-1}(\varphi, \vartheta) \cdot E' \quad (A.6)$$

Using these identities, one can calculate for the second-order nonlinear polarization

$$P_i = \varepsilon_0 \, \chi_{ijk}^{(2)} \, E_j \, E_k$$

$$\Leftrightarrow \qquad T_{il}^{-1} \, P_l' = \varepsilon_0 \, \chi_{ijk}^{(2)} \, T_{jm}^{-1} \, E_m' \, T_{kn}^{-1} \, E_n'$$

$$\Leftrightarrow \qquad P_l' = \varepsilon_0 \, T_{li} \, T_{jm}^{-1} \, T_{kn}^{-1} \, \chi_{ijk}^{(2)} \, E_m' \, E_n'$$

$$\Leftrightarrow \qquad P_l' = \varepsilon_0 \, \chi_{lmn}'^{(2)} \, E_m' \, E_n' \tag{A.7}$$

where $\chi_{lmn}'^{(2)}$ is the effective second-order nonlinear susceptibility tensor in beam coordinates:

$$\chi_{lmn}'^{(2)} = T_{li} \, T_{jm}^{-1} \, T_{kn}^{-1} \, \chi_{ijk}^{(2)} \tag{A.8}$$

The same rules apply for the third order tensor

$$\chi_{mnpq}'^{(3)} = T_{mi} \, T_{jn}^{-1} \, T_{kp}^{-1} \, T_{lq}^{-1} \, \chi_{ijkl}^{(3)} \tag{A.9}$$

or any higher order tensors.

A.3 Influence of Pointing on the Timing Jitter

In Sect. 3.3.3 it was mentioned that the simple calculation by Klingebiel et al. (compare Fig. 4 in [2]) regarding the impact of pointing on the timing jitter is in our opinion not entirely correct. The missing aspect in their calculation is the non-collinear angle between the intersecting beams in the cross-correlation setup used to measure the jitter (cf. Fig. 3.10) as well as in the OPA stages. In the following we will show that in general this aspect is important to quantify the temporal effect of pointing. It will, however, turn out that even though the effect is two orders of magnitude stronger than expected before, it is not the dominant source of jitter at our system.

In Fig. A.1 once more the cross-correlation scheme from Fig. 3.10 is shown,[1] but this time with an additional pointing source P (e.g. a vibrating mirror or air fluctuations) in the optical path of the pump beam that deflects the beam randomly from shot to shot. As a consequence, also the point of intersection with the Femtopower reference is changing. If we assume that the pointing source only affects the beam *direction* and does not introduce any time delay by itself, an undeflected pump pulse shot arrives at B at the same time as a deflected shot arrives at C (cf. Fig. A.1). To reach A, i.e. the point of intersection of the deflected shot with the Femtopower beam, an additional travel time

$$\Delta t_{\mathrm{p}} = \frac{\overline{CA}}{v_{\mathrm{gr,p}}} \tag{A.10}$$

---

[1] Note that the actual setup is slightly more complex since pump and Femtopower beam are focused, there is refraction at the crystal surfaces, etc. However, these simplifications do not affect the validity of the considerations.

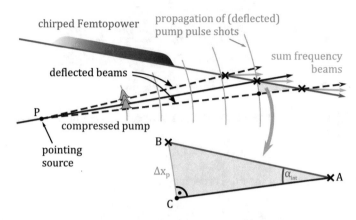

Fig. A.1 Illustration of the impact of beam pointing on the timing jitter measurement with the cross-correlation technique. Due to the non-collinear angle α_{int}, the deflected pump beams intersect with the Femtopower reference at different points marked by crosses. As the travel time to reach these points is different for pump and reference (for details see main text), the relative delay between the pulses (=jitter) depends on the point of intersection, i.e. on pointing

is required where $\overline{CA}$ is the distance between C and A and $v_{\text{gr,p}}$ is the group velocity of the pump pulses in the crystal:

$$v_{\text{gr,p}} = \frac{c}{n(\lambda_{\text{p}}) - \lambda_{\text{p}} \frac{\partial n}{\partial \lambda}\big|_{\lambda_{\text{p}}}} \tag{A.11}$$

with the vacuum speed of light c and the ordinary refractive index n of BBO. Correspondingly, it takes the time

$$\Delta t_{\text{FP}} = \frac{\overline{BA}}{v_{\text{gr,FP}}} \tag{A.12}$$

for the Femtopower pulses to travel from B to A, where $v_{\text{gr,FP}}$ is the group velocity for these pulses.

Let us assume that the undeflected pump and Femtopower pulses arrive simultaneously at B. Then it follows from the previous considerations that for a deflected pump beam, the pulses do not reach A at the same time due to the different travel times. A relative temporal delay Δt between the pulses is created:

$$\Delta t = \left|\Delta t_{\text{p}} - \Delta t_{\text{FP}}\right| = \left|\frac{\overline{CA}}{v_{\text{gr,p}}} - \frac{\overline{BA}}{v_{\text{gr,FP}}}\right| \tag{A.13}$$

By trigonometric rules one can relate the distances $\overline{CA}$ and $\overline{BA}$ to the non-collinear angle α_{int} and to the distance between B and C which we denote as Δx_{p}:

$$\overline{CA} = \frac{\Delta x_p}{\tan \alpha_{int}} \tag{A.14}$$

$$\overline{BA} = \frac{\Delta x_p}{\sin \alpha_{int}} \tag{A.15}$$

The quantity Δx_p can be understood as the spatial fluctuations of the pump beam at the position of the crystal due to pointing. Since there is also pointing of the Femtopower beam resulting in additional spatial fluctuations Δx_{FP}, we merge Δx_p and Δx_{FP} in one parameter (assuming statistically uncorrelated fluctuations):

$$\Delta x = \sqrt{\Delta x_p^2 + \Delta x_{FP}^2} \tag{A.16}$$

Combining Eqs. A.13 to A.16, one obtains an expression for the timing jitter due to pointing:

$$\Delta t = \Delta x \left| \frac{1}{v_{gr,p}} \frac{1}{\tan \alpha_{int}} - \frac{1}{v_{gr,FP}} \frac{1}{\sin \alpha_{int}} \right| \tag{A.17}$$

At a non-collinear angle $\alpha_{int} = 2.3°$ in our setup and measured spatial fluctuations of $\Delta x_p = 25\,\mu m$ (RMS of several hundred shots) and $\Delta x_{FP} = 19\,\mu m$, this would result in a timing jitter of $\Delta t = 28.9\,fs$.

It is important to note, however, that Eq. (A.17) is only true if the two intersecting beams spatio-temporally overlap entirely inside the crystal. This is shown in Fig. A.2a for small (compared to Δx) beam diameters and in Fig. A.2b for large diameters. While for small beams the intersection is point-like, for large beams there is an area of overlap.[2] In the latter case, the effective delay between the pulses is determined by averaging over all relative delays inside this area or simply by taking the delay at the center of mass. As can be seen these center points in Fig. A.2b correspond to the points of intersection in Fig. A.2a which implies that the effective delay can again be calculated with Eq. (A.17).

In contrast, if the crystal thickness d is smaller than the longitudinal pointing

$$d < \overline{BA} = \frac{\Delta x_p}{\sin \alpha_{int}} \tag{A.18}$$

as shown in Fig. A.2c, some of the deflected pulses might not overlap anymore with the Femtopower pulses inside the crystal and hence generate no sum-frequency signal. Thus, the maximum time delay that can be measured in this case with the cross-correlation setup is

---

[2]Note that this geometrical approach is just a first-order approximation and that the overlap in fact depends on many parameters such as the exact beam profiles, the angle between the pulse fronts or walk-off.

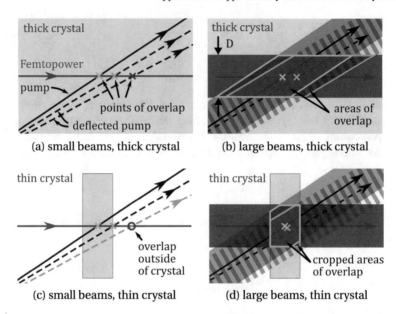

(a) small beams, thick crystal

(b) large beams, thick crystal

(c) small beams, thin crystal

(d) large beams, thin crystal

Fig. A.2 Spatial overlap of pump and Femtopower pulses in a nonlinear crystal for three different scenarios: **a** If beams are small, the overlap can be considered to be point-like. **b** If beams are large, there is an area of overlap. The centers of mass of the areas correspond to the points in (**a**). **c** If the crystal is thin, some of the deflected pulses do not overlap with the Femtopower inside the crystal and hence produce no sum-frequency signal. **d** If beams are large and the crystal is thin, the effective overlap areas shrink to the crystal dimension shifting the centers of mass

$$\Delta t_{\max} = \frac{d}{2} \left| \frac{\cos \alpha_{\mathrm{int}}}{v_{\mathrm{gr,p}}} - \frac{1}{v_{\mathrm{gr,FP}}} \right|. \tag{A.19}$$

At a crystal thickness $d = 500\,\mu\mathrm{m}$ in our setup, the maximum observable pointing-induced jitter is therefore limited to $\Delta t_{\max} = 9.2\,\mathrm{fs}$.

For large beam diameters D, the effective delay between the pulses is once more reduced by an effect shown in Fig. A.2d: if the crystal is smaller than the longitudinal dimension of the overlap area, i.e.

$$d < D \left(\frac{1}{\tan \alpha_{\mathrm{int}}} + \frac{1}{\sin \alpha_{\mathrm{int}}} \right) \overset{\alpha_{\mathrm{int}} \ll 1}{\approx} \frac{2\,D}{\alpha_{\mathrm{int}}} \tag{A.20}$$

the effective overlap area shrinks to the crystal dimension. As a consequence, the center points of these areas are located closer to each other and therefore the effective delay between the pulses is smaller.

Analytically, this effect is difficult to quantify, therefore we performed a quick quasi-3D numerical study with Gaussian-shaped beams. We calculated the regions of overlap for deflected and undeflected beams, cropped them by the crystal thickness and determined the respective centers of mass. Using the experimentally determined beam sizes in the cross-correlation setup of $D = 100\,\mu\text{m}$ for both beams, one obtains an effective maximal time delay of $\Delta t_{\text{max,eff}} = 1.5\,\text{fs}$. This value finally is the jitter that is expected to be measured in the presence of the specified pointing.

Due to the larger beam sizes at the OPA stages (of the order of millimeters), the pointing-induced jitter between pump and signal is significantly lower than in the cross-correlations setup (assuming identical pointing) and only of the order of 10–500 as. Interestingly, this implies that the timing stabilization based on the measured jitter value can even increase the actual jitter in the OPA stages. But since the measured total jitter is about 80 fs (cf. Sect. 3.3.3), it is obvious that pointing does not represent the dominant source of jitter and therefore can be neglected.

However, as the effect is two orders of magnitude larger than the previously calculated value of 17 as [2], the presented considerations might be helpful for the proper design of jitter-stabilization systems for applications where other jitter sources have already been suppressed and where fs-precision is required, e.g. for the synthesis of few-cycle pulses.

A.4 Additional Calculations for the Idler Generation Scheme

A.4.1 Derivation of the Wavelength-Dependent Idler Exit Angle

In the following, Eq. (4.1) from Sect. 4.1 will be derived. To do so, we shall recall two identities: Energy conservation requires that the idler frequency is

$$\omega_i = \omega_p - \omega_s \qquad (A.21)$$

and momentum conservation in a non-collinear scheme requires that the idler wavenumber is

$$k_{i,\text{pm}} = \sqrt{k_p^2 + k_s^2 - 2\,k_p\,k_s\,\cos\alpha_{\text{int}}} \qquad (A.22)$$

where the subscript "$_{\text{pm}}$" indicates that this equation holds only in the phase-matched case. The geometrical equivalent to Eq. (A.22) is depicted in Fig. A.3a.

Applying the sine rule yields

$$\frac{k_s}{\sin(\beta_{\text{int}})} = \frac{k_{i,\text{pm}}}{\sin(\alpha_{\text{int}})} \qquad (A.23)$$

$$\Rightarrow \quad \sin(\beta_{\text{int}}) = \frac{k_s}{k_{i,\text{pm}}}\sin(\alpha_{\text{int}}) \qquad (A.24)$$

$$= \frac{\sin(\alpha_{\text{int}})}{\sqrt{1 + k_p^2/k_s^2 - 2\,k_p/k_s\,\cos(\alpha_{\text{int}})}} \qquad (A.25)$$

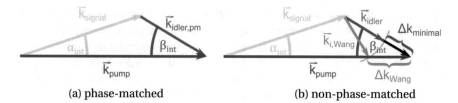

(a) phase-matched   (b) non-phase-matched

Fig. A.3 Depiction of the wavevectors of a non-collinear OPA in the **a** phase-matched and **b** non-phase-matched case. In both cases k_{idler} points towards the tip of the pump wavevector to minimize the wavevector mismatch Δk

and finally

$$\beta_{ext} = \sin^{-1}\left(n_i \cdot \sin\left(\beta_{int}\right)\right)$$
$$= \sin^{-1}\left(\frac{n_i \sin\left(\alpha_{int}\right)}{\sqrt{1 + k_p^2/k_s^2 - 2\,k_p/k_s \cos\left(\alpha_{int}\right)}}\right). \tag{A.26}$$

Note that in the realistic case of imperfect phase matching it is

$$k_i = \frac{\omega_i\, n_i}{c} \neq k_{i,pm} \tag{A.27}$$

so momentum conservation is violated and Eq. (A.22) does not hold anymore. Geometrically, this implies that the length of k_i is changed and the triangle of k_p, k_s and k_i is not closed anymore as shown in Fig. A.3b. However, the "preferred" direction of k_i is the one that minimizes Δk and therefore the angle β_{int} is the same as in Fig. A.3a. For comparison, the vector $k_{i,Wang}$ as calculated in [3] is shown that has the same length as k_i but is pointing in a different direction. Since this direction does not minimize Δk, the angle β_{int} reported in Eq.(1) of [3] is in our opinion not correct and Eq. (A.25) should be used instead.

A.4.2 Derivation of the Idler Angles After Compensation

Here we briefly derive Eq. (4.2), i.e. the wavelength dependent output angle of the idler beam $\vartheta_{out}(\lambda_i)$ after having passed the compensation scheme consisting of telescope and grating. As already mentioned in Sect. 4.1.2, the bundle of idler rays is centered around the optical axis of the telescope to reduce aberrations:

$$\beta_{tel}(\lambda_i) = \beta_{ext}(\lambda_i) - \beta_{ext}(\lambda_{i,central}) \tag{A.28}$$

For the telescope (or any other) imaging system with magnification $M = f_2/f_1$ holds:

$$\tan\left(\beta'_{\text{tel}}(\lambda_i)\right) = \frac{\tan\left(\beta_{\text{tel}}(\lambda_i)\right)}{M} \tag{A.29}$$

Note that neglecting the tangent as has been done in [3] is not valid since the angles in the idler generation setup are not small.

The well known equation for the first diffraction order from a grating with line separation d states [4]:

$$\frac{\lambda}{d} = \sin\vartheta_{\text{in}} + \sin\vartheta_{\text{out}} \tag{A.30}$$

Hence, with the input angle $\vartheta_{\text{in}}(\lambda_i) = \beta'_{\text{tel}}(\lambda_i) + \vartheta_{\text{gr}}$ (compare Fig. 4.2) one obtains for the output angle

$$\vartheta_{\text{out}}(\lambda_i) = \sin^{-1}\left[\frac{\lambda_i}{d} - \sin\left(\tan^{-1}\left(\frac{\tan\left(\beta_{\text{tel}}(\lambda_i)\right)}{M}\right) + \vartheta_{\text{gr}}\right)\right] \tag{A.31}$$

which is the equation (4.2) to be derived.

A.5 Contrast Deterioration by a Liquid-Crystal Spatial Light Modulator

In this section, the potential pulse distortions introduced by an LC-SLM due to its intrinsic pixelation will be discussed. For the case study we will use the specifications of the device implemented into our system as listed in Table A.2. From the spatial chirp created by the zero-dispersion stretcher described in Sect. 5.2.3, one can infer an effective number of 541 stripes that are covered by the examined spectral components at 650–1500 nm.

In general, there are two independent effects from pixelation that can deteriorate the temporal contrast: First, for technical reasons there are small gaps ($3\,\mu$m wide) between the individual stripes of the liquid-crystal array. Being opaque, these gaps effectively block spectral components of the spatially chirped pulses passing

Table A.2 Specifications of the implemented liquid-crystal spatial light modulator

Device type	Jenoptik SLM-S640
Active area	64 mm × 10 mm
Number of stripes	640
Fill factor	97 %
Wavelength range	430–1600 nm
Maximal phase shift	$\sim 7\pi$ @ 430 nm
	$\sim 2\pi$ @ 1600 nm

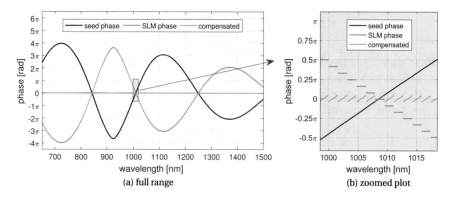

Fig. A.4 Simulation: arbitrary spectral phase of the seed pulses and compensation by the LC-SLM. Due to the pixelation of the device, the introduced SLM phase is stepped and hence the phase of the compressed pulse is represented by a saw-tooth function

the array. The resulting modulation of the spectrum corresponds to artifacts in the temporal waveform.

Second, the intended task of the LC-SLM is to adjust the spectral phase of the transmitted pulses. This is achieved by applying different voltages to the individual stripes of the LC-SLM which results in a rotation of the liquid crystals and an according phase shift for e-polarized light. As this phase shift, however, is constant within one stripe, there are discrete phase jumps at the stripe borders. For the pulses passing the liquid-crystal array this leads to a stepped spectral phase which can again affect the temporal contrast. In the following we will try to quantify the influence of these effects.

As the input phase of the seed pulses, an arbitrary curve (see Fig. A.4) is chosen that is at the limit of what can be compensated by the LC-SLM. The unavoidable discretization of the introduced phase and the resulting saw-tooth-like output phase is shown in the magnified plot in Fig. A.4b.

The distortions of the spectral amplitude are displayed in Fig. A.5: Fig. A.5a shows the input seed spectrum without pixelation. Figure A.5b shows the distorted spectrum for the hypothetic case of perfect spatial separation of wavelengths in the Fourier plane, i.e. at the LC-SLM. This would, however, imply that all wavelengths are focused by the zero-dispersion stretcher setup to infinitely small spots. The more realistic case of a finite focus diameter of ∼70 μm FWHM (measured in our system) is shown in Fig. A.5c: the transmitted spectrum corresponds to a convolution of the Gaussian focus with the discrete structure of the liquid-crystal array.

The same applies for the spectral phase jumps from Fig. A.4: they are smoothed by the spatial averaging of the real beam with just a small remaining modulation. In Fig. A.6, the temporal contrast for the different cases is shown where several important observations can be made: Rather than creating short, isolated pre- and post-pulses, the simulated distortions due to pixelation result in picosecond-long structured side-wings. In fact, this is a consequence of the angular dispersion intro-

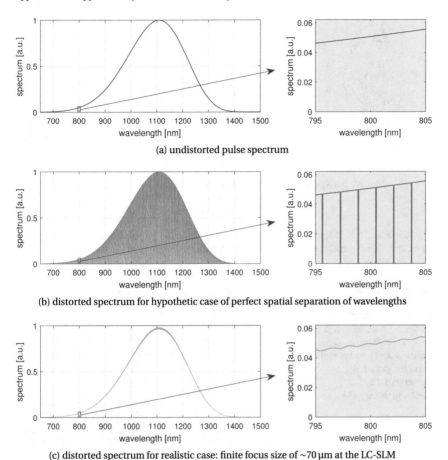

Fig. A.5 Simulated distortion of the spectral amplitude due to the pixelation of the LC-SLM. Left column shows distortions of the full spectral range, right column shows zoomed in plots for better visibility

duced by the grating in the zero-dispersion-stretcher, which generates a spatial chirp linear in wavelength (and *not* in frequency). For the realistic case in Fig. A.6c, the additional temporal features reach a relative intensity of up to $\sim 3 \times 10^{-7}$ and are confined to the region -4 ps to 1 ps (as well as 1 ps to 4 ps). This position of the side wings is determined by the number of effectively used stripes of the LC-SLM: if for example the otherwise identical Jenoptik SLM-S320 with only half the number of total stripes was used, the side wings would be a factor of two closer to the main peak. Finally, the direct comparison of black and colored curves in the plots reveals that phase distortions can boost the adverse effect of amplitude distortions on the time structure of the pulse by more than an order of magnitude.

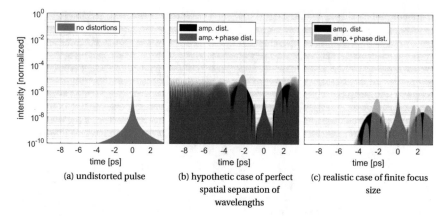

(a) undistorted pulse

(b) hypothetic case of perfect spatial separation of wavelengths

(c) realistic case of finite focus size

Fig. A.6 Simulated temporal contrast of **a** ideal, undistorted pulses, **b** distorted pulses in the hypothetic case of perfect separation of wavelengths in the Fourier plane and **c** distorted pulses in the realistic case

Fig. A.7 Contrast measurement of the seed pulses with an XFROG. Within the dynamic range of the setup no features are visible in the temporal region that could potentially by compromised by the pixelation of the LC-SLM

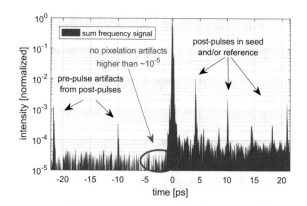

To test these predictions experimentally, we performed an XFROG measurement of the partially compressed seed pulses after passing the zero-dispersion-stretcher, the LC-SLM and the OPA vacuum system. The gate pulses were provided by the reference output channel of the Femtopower (25 fs, 20 µJ). The result of the measurement, i.e. the spectrally integrated sum-frequency signal is shown in Fig. A.7. Due to the low seed pulse energy of less than 1 µJ at the diagnostic setup and the limited sensitivity of the spectrometer that detects the generated sum frequencies, the temporal contrast could only be measured with a dynamic range of $\sim 10^{-5}$. At this level, however, no temporal features could be detected in the region of interest at -4 ps to -1 ps which supports the theoretical predictions.

In summary, two aspects in the theoretical and experimental data are crucial for the usability of the examined LC-SLM in the PFS system: First, due to the pixel resolution of the device, the created temporal artifacts are located outside the pump pulse window of about ±0.5 ps and will thus not be amplified during OPA. And second, the intensity of these artifacts relative to the main peak is estimated to be $<10^{-6}$. Hence, at a parametric gain of more than 10^4, the expected contrast of the amplified signal pulses is sufficiently good at $>10^{10}$.

References

1. R.W. Boyd, *Nonlinear Optics*, Academic Press, 2008.
2. S. Klingebiel, I. Ahmad, C. Wandt, C. Skrobol, S. A. Trushin, Z. Major, F. Krausz, and S. Karsch, Experimental and theoretical investigation of timing jitter inside a stretcher-compressor setup. Opt. Express **20**, 3443–3455, (2012). https://doi.org/10.1364/OE.20.003443.
3. T.J. Wang, Z. Major, I. Ahmad, S. Trushin, F. Krausz, S. Karsch, Ultrabroadband near-infrared pulse generation by noncollinear OPA with angular dispersion compensation. Appl. Phys. B Lasers Opt. **100**, 207–214, (2010). https://doi.org/10.1007/s00340-009-3800-9.
4. D. Meschede, *Optik, Licht und Laser* (Vieweg+Teubner Verlag, 2008). https://doi.org/10.1007/978-3-8348-9288-1.

Appendix B
Data Archiving

The experimental raw data, the corresponding evaluation files and the resulting figures can be found on the data-archive server of the Laboratory for Attosecond Physics at:

  //AFS/ipp-garching.mpg.de/mpq/lap/publication_archive

The files are organized in accordance with the structure of the thesis: For each figure, there is a dedicated folder containing the raw data, Matlab scripts for data analysis or simulations, and the figure. If necessary, a readme.txt file provides further explanations and instructions.

© Springer International Publishing AG, part of Springer Nature 2018   165
A. Kessel, *Generation and Parametric Amplification of Few-Cycle*
Light Pulses at Relativistic Intensities, Springer Theses,
https://doi.org/10.1007/978-3-319-92843-2

Printed in the United States
By Bookmasters